Projections and Origins

collected writings
of
Brian Adams

Edited by Roger Hellyer
and Chris Higley

London
The Charles Close Society
2006

Published by the
Charles Close Society
for the Study of
Ordnance Survey Maps

c/o The Map Library
British Library
96 Euston Road
London, NW1 2DB

www/charlesclosesociety.org.uk

ISBN 1 870598 26 1

Printed by Joshua Horgan, Oxford

Contents

Preface

Brian Adams's consuming passion was the mathematical content of maps – the baselines and triangulation systems, the projections, the origins and meridians, the organisation of sheet lines and ultimately the mapping drawn therein. He had a rare ability of making these seemingly impenetrable areas of map making comprehensible to the layman, and he had a facility of writing about these subjects in a way that engages the reader who never thought him or herself capable of being the least bit interested in the subject.

In this collection we have brought together several of Brian's writings, both on these subjects and others that interested him, mostly taken from the further reaches of *Sheetlines*. It is hoped that in this new format they will become much more accessible and useful a resource to those who wish to pursue further the themes of Brian's own research interests. Other published work to have come to our notice is listed in the bibliography on page 110.

Brian turned to Ordnance Survey mapping as the central theme of his research interests after 1980, the year he retired from 35 years service with Hydrographic Department. There is ample evidence of his interest in the subject during those earlier years, but he had little opportunity to write about it. In his chosen area of academe, Brian was not just one expert, he was the expert. His investigations revealed just how incomplete were the Ordnance Survey's own records of County Series mapping, which had after all formed some 90% or more of its output since 1833. Many county origins were unknown, and the actual co-ordinates of some which were known he proved to have been inaccurately recorded. Brian's great achievement was painstakingly to discover how many county origins there actually were (remarkably an unknown quantity in itself), to identify their names, to confirm and where necessary compute their precise locations, and to calculate the sheet line co-ordinates derived from them. Brian refused to release any of this information until he had satisfied himself completely of its accuracy. Enough detail has emerged from his research papers to convince us that he was so satisfied, and thus these data, fundamental to an understanding of the entire system of County Series mapping, in both Great Britain and Ireland, appear here in print for the first time.

As Yolande Hodson put it in *Popular maps*, "anyone familiar with Adams's work will know that [the figures] have been rigorously double-checked" – thus any errors that readers stumble upon below are not the fault of Brian, but a failure of the editors and proof reading.

What did he leave undone? Two unfinished research projects come to mind. The first was his investigation into the construction of one-inch Old Series mapping in the area below the Preston to Hull line, west of the Greenwich and east of the 3° West blocks of sheets. The second was to discover the origins of those large scale town plans not laid out on County Series meridians, usually euphemistically described as being on "local sheet lines". As Brian once wrote, albeit in another context, "it is surely obvious, once the user's attention has been drawn to it, that [rectangular] sheet lines are parallel and perpendicular to something, and at least some of them would be interested to know what". Perhaps someone may take up the challenge.

The Charles Close Society would like to express its gratitude to the beneficiaries of Brian Adams's estate for their support in the publication of this volume, and for their generosity in releasing Brian's research papers, together with any maps and books that were relevant, to the Charles Close Society Archives, held in the Map Department of Cambridge University Library. A catalogue of the materials available in this unique resource appears on the society's website. We would like also thank David Archer for kindly allowing essays which Brian contributed to two of his publications to appear in these pages.

iv

Foreword [1]

Maps come in an extraordinarily wide variety, ranging from a how-to-get-there-from-here sketch to the fully cmpetent professional product, and may be manuscript, rapid copied, printed, or computer-generated; they may be individual, scattered through a book or magazine, in bulk in a traditional atlas by the hundred, or the hundred-thousand in a computer memory; and they come in wide varieties of size, of scale, and of content, not forgetting pure fiction such as Treasure Island. But in any consideration of total cartography the expression **Maps and Charts** must always be used, for depictions of Water and Air have to take their places along with those of Earth and occasionally, regrettably, Fire.

However, the maps included in the present work are distinguished by being the products of a particular group of national mapping agencies, originally one but now metamorphosed into three, the Ordnance Surveys of the British Isles, and comprise all the smaller scale maps produced by those agencies. The Ordnance Survey of Great Britain and Ireland dated its foundation from 1791, although that name only came along later, and its first operations were based on Major-General William Roy's triangulation of south-east England, carried out in 1784-88 as part of an Anglo-French operation to connect the observatories of Paris and Greenwich. Much of this was re-observed, and it was then massively extended to complete the first Principal Triangulation of the British Isles. This framework formed the basis for topographical surveys which the Survey used to construct a national map series, initially of England and Wales, on the virtually self-selecting and highly convenient scale of one inch to one statute mile, the one-inch map that became familiar as *the* Ordnance Map to much of the general public.

Before one-inch cover had been completed, the Ordnance Survey ventured into both smaller and larger scale maps. The latter were mainly used by particular professions, and totalled over 100,000 sheets by the turn of the century; they have been thoroughly described and indexed in several publications of David Archer and of the Charles Close Society.[2] Meanwhile it is the purpose of the present volume to provide indexes and identify all the smaller scale Ordnance Survey maps comprehensively for the first time, the range of scales being from 1:50,000, metric successor to the one-inch, down to 1:4,000,000, with descriptions of a few larger scale maps which are relevant to the smaller scale story. The remainder of this Foreword is devoted to an outline of the construction methods of Ordnance Survey maps and map series, and associated matters. These are things that the average map user takes entirely for granted, but a pause for thought will tell that they are vital to the production of maps, and without them there would be no maps to look at.

Construction of Ordnance Survey maps

The reader will undoubtedly be aware that the representation of the curved surface of the Earth by a flat map requires the use of a **Map Projection**, and has probably seen some rather frightening diagrams of various projections at the front of an atlas. But whilst those projections are mainly used for very small-scale maps and need not concern the users of

[1] Reprinted, by kind permission of the publisher, from Roger Hellyer, *Ordnance Survey small-scale maps indexes : 1801-1998*, Kerry: David Archer, 1999, ISBN 0 9517579 54.

[2] Richard Oliver, *Ordnance Survey maps a concise guide for historians*, second edition, revised and expanded, London: Charles Close Society, 2005; *Indexes to the 1/2500 and 6-inch scale maps : England and Wales*, Kerry: David Archer, 2002; *Indexes to the 1/2500 and 6-inch scale maps : Scotland*, Kerry: David Archer, 1993.

Ordnance Survey maps, a projection is still required for the construction of these maps as well as for laying down the actual topographical surveys to provide the map detail.

Most recent Ordnance Survey map series, but only a few of the earlier ones, have a statement in the marginal material of the map projection used. However, a prominent feature of the modern maps is the system of squares, the National Grid, printed across the face of the map and numbered in sequence. This is presented to the ordinary map user simply as a reference system, but strictly speaking this usage is only a by-product of the grid's fundamental purpose. What are known as the full National Grid co-ordinates, the Eastings and Northings *including* the small figures which the user is instructed to ignore, are in fact the *x,y* rectangular co-ordinates needed for constructing the map on the relevant (Transverse Mercator) projection, and derive from formulae expressing the co-ordinates in terms of latitude and longitude. These formulae are quite complex due to the need to allow for the ellipsoidal shape of the Earth in constructing the large-scale maps which the Ordnance Survey produces, as well as in rendering its accurate surveys into plane, paper-flat, terms.

The user whose interest has been aroused by the stated use of the Transverse Mercator Projection for the later series of Ordnance Survey maps may then feel frustrated by the total absence of any similar attribution on most earlier series. But wait, are not the latter's sheet line systems also rectangular?, and do not some have reference squares printed across them? Yes indeed, and all these lines are *x,y* or Easting and Northing co-ordinate lines on the particular projections involved, although not numbered with their co-ordinate values. Strictly speaking, the origins of the Survey's two main projections lie on the Equator, but for convenience algebraic origins (co-ordinate zeros) are chosen within the area of mapping, wherefrom the co-ordinates may be negative or positive, or in other words may be west or south as well as east or north.

Thus, for example, the sheet lines of one-inch *Popular Edition* Sheet 75 *Ely* are, in cyclic order from the northern neat line, 236,180 feet south, 709,110 feet east, 331,220 feet south, 566,550 feet east, from the projection origin of Delamere (of which more below). The map is crossed by a net of two-mile squares with an alpha-numeric reference system, but the lines forming these squares are co-ordinate lines at 10,560 feet intervals. The reason that National Grid co-ordinates are all east and north is explained below.

A detailed study of all the projections used for the construction of the Ordnance Surveys' "ten-mile" maps (scales 1:633,600 and 1:625,000) has been made by myself,[3] and is to be found on page 5. Incorporated in that study are all the projections which have been used for the Survey's regular map series on scales larger than 1:625,000 that are included in the present work. No similar review is available covering the projections used for Ordnance Survey maps at scales smaller than 1:633,600, but where known, details are included in the individual map or series sections; in the absence of any other information, the relevant items are taken from the title or other material on the face of the map, unless my own professional experience suggests any reservations thereon, as such annotations are not always totally to be trusted.

In a number of the particularly small-scale series the sheet lines are not rectangular, but are formed by portions of the **Graticule** comprising parallels of latitude and meridians of longitude as depicted on the relevant projection. On some projections, usually described as *Modified*, the meridians are drawn straight instead of very slightly curved as they would be on the true projection.

---

[3]   originally published in Roger Hellyer, *The 'ten-mile' maps of the Ordnance Surveys*, London: Charles Close Society, 1992.

In the body of this volume many of the projection references are given in standardised form, and to avoid confusion the various projections are referred to by their accepted names as given in the Royal Society *Glossary of Technical Terms in Cartography*. Meanwhile, the reader who wishes for extended technical details of map projections is referred to Maling.[4] The simple reference "National Grid series" or occasionally "National Grid map" signifies in total that the maps are constructed on the Transverse Mercator Projection, specifically the 1936 national projection with its origin at 49°N, 2°W, that the sheet lines are formed by co-ordinate lines of the national projection (i.e. National Grid lines), and that the National Grid is printed across the face of the maps.

The national projection originated in 1929 in the days of imperial units, and pre-war maps constructed on it carried the National Yard Grid; whilst this grid and the National Grid in its early days were subject to some variation in nomenclature, they are given the standard names herein for the avoidance of doubt. Maps originally published with the National Yard Grid and later converted to the National Grid will be found to have the latter's sheet line co-ordinates in very non-round figures.

"Irish National Grid series" or "Irish National Grid map" signify that the projection is the Transverse Mercator Projection of Ireland with its origin in 53° 30' N, 8° W, that the maps have Irish National Grid sheet lines, and have that grid across the face of the maps.

For over eighty years prior to the adoption of the 1929 projection the regular series of Ordnance Survey maps of England and Wales, and latterly of Great Britain, were constructed on Cassini's Projection on the origin of Delamere. This was an early primary triangulation station, known originally as Delamere Forest, situated in Cheshire in position 53° 13' 17".274 North, 2° 41' 03".562 West.

The large-scale plans in both Ireland and Scotland had been constructed on Cassini's Projection, on a county-by-county basis, but when the Ordnance Survey came to prepare the one-inch series of these countries they adopted Bonne's Projection. No firm reason is known for that decision, although Bonne's had some vogue at the time, but it involved an extra stage in the reduction of the large-scale material to provide the detail for the one-inch mapping. This was partly due to the curvature of the parallels on Bonne's Projection being generally noticeably different to that on Cassini's. For the first time these two series used origins which were precise intersections of latitude and longitude, the Irish being in 53° 30' N, 8° W, and the Scottish in 57° 30' N, 4° W.

In due course the Scottish one-inch and smaller scale mapping were republished on Cassini's Projection on the origin of Delamere, the better to join up with English mapping. Prior to this, some of the smaller scale Great Britain series had combined maps on the two projections with varying degrees of success, and I have dubbed these (with hopeful decorousness) as mixed parentage mapping.

The *International Map of the World on the Millionth Scale,* to give it its proper title in the English-speaking world, has its sheets constructed on their own projection, devised by the map's originating committee and a form of Modified Polyconic Projection. This is referred to in the body of this work as the *IMW Projection.*

True conical projections are calculated with reference to a central parallel of latitude, and have no central meridian as such. However, they may well have an initial meridian which is used for the calculations to lay out the projection, and possibly for the zero true easting line

---

[4] D H Maling, *Coordinate systems and map projections*, second edition, Oxford: Pergammon Press, 1991.

of a military grid. *Minimum Error Conical Projection* does not refer to a specific projection, but to a type of projection in which specified classes of error are reduced to a minimum when meaned across the map. The process may be carried out conventionally or mathematically, and some projections are more "minimum error" than others. It is, however, probable that those Ordnance Survey maps quoted as Minimum Error Conical Projection are all on the same basic projection, with difference due to latitude.

As I mentioned earlier, National Grid co-ordinates are fundamentally the dimensional construction co-ordinates of the 1936 national projection, but are also utilised as a reference system. However, it is from the latter usage that the term **Grid** derives, since in cartography the word originally denoted an arbitrary system of squares printed on the face of a map and used for military referencing. Arbitrary systems were soon largely replaced by survey-related systems, where the grids appeared as true squares on the maps they were initially applied to. But the grids had to be printed on other maps to provide unique referencing, and in such cases might not be precisely true squares (although not necessarily visibly out of true), and would very often lie at an angle to the new sheet lines.

When a grid is stated to be on a certain projection, this is its original map projection upon which it forms absolutely true squares and for which it also provides the construction co-ordinates. However, to eliminate a potent source of error in referencing, grid co-ordinates are usually measured from a **False Origin** lying to the south-west of the mapped area, rendering all co-ordinates positive, that is east and north only. Thus the False Origin of the National Grid lies west by south from the Isles of Scilly and south-west of the whole of Great Britain, whereas the **True Origin** of the national projection lies to the south of Dorset.

Derived series

Many Ordnance Survey map series are derived directly or indirectly from series on other scales, generally involving a reduction of scale, but occasionally an increase for some particular purpose; the reader should appreciate that in the latter instance there is no gain of accuracy. Sometimes a change of projection may be deliberately introduced along with the change of scale, but otherwise two different effects are possible. Where maps have been constructed on one of the transverse cylindrical projections, Cassini's or Transverse Mercator, or on Bonne's Projection, then any amount of amalgamation or sub-division of sheets results in a map on the same original projection. But in other cases, with the International Map of the World being especially involved, the result from manipulation of sheets is a map which as an entity is on no projection at all. Such a map can then only truly be described by its method of derivation from the original.

Parsons Green
London SW

Projections of the Ordnance Survey ten-mile maps

"Begin at the beginning, and go on till you come to the end; then stop." This simple rule, enunciated gravely by the King of Hearts, may perhaps seem unusually straightforward for the Wonderland of Alice; I find it altogether inappropriate for the wonderland of the Ordnance Survey ten-mile map projections. Instead I propose to start in the middle, go on to the end, then backwards to the beginning, and then across the sea to Ireland; and somewhere I need to fit in a projection whose use was very largely mythical. Reader, take heed! But first:-

1. Ordnance Survey fundamental elements

The projections of all Ordnance Survey maps of Great Britain, of regular series commenced after 1830 and on scales of 1:633,600 and larger, are calculated on the Airy spheroid (Airy's figure of the Earth) defined in terms of the foot of Bar O_1. All such projections prior to the adoption of the National Projection had a scale factor of unity. Maps of Ireland on scales from 1:63,360 to 1:253,440 inclusive, and all scales on the Transverse Mercator Projection of Ireland until 1965, are also on the Airy spheroid, and all these have a scale factor of unity.

The reader will observe that several of the ten-mile maps fall outside the above-mentioned ranges, either of date or in Ireland of scale, and further details of these maps will be found in the relevant sections following. All subsequent references to feet, including yards proportionally, are to feet of Bar O_1.

2. The properly projected series

1. Three-sheet series, 1926

This was the first series of ten-mile maps of Great Britain to be properly projected throughout; it was constructed on Cassini's Projection on the origin of Delamere, latitude 53° 13' 17".274 N, longitude 2° 41' 03".562 W, the unit of calculation being the foot. This specific projection had originally been adopted for the northern sheets of the one-inch Old Series maps of England & Wales, followed by the one-inch New Series and quarter-inch and half-inch maps of those countries. But it was only after the first world war that its use was extended to maps of Scotland, on the one-inch and quarter-inch scales, thereby facilitating its adoption for an all Great Britain smaller scale series.

I have described elsewhere[1] how the two mile squares on one-inch Popular Edition maps were formed by rectangular co-ordinate lines of this same Cassini projection, and so also are the squares of the two inch, twenty scale mile, reference grid on this ten-mile series. Thus each two inch interval represents 105,600 feet on the projected spheroid, north-south distances on the ground usually being slightly less due to the effects of the projection. According to Ordnance Survey records[2] the co-ordinates of the sheet lines of the series are as follows:-

Common west neat line	1,006,890 feet west
Common east neat line	999,510 feet east

---

[1] '198 years and 153 meridians, 152 defunct', see page 53.
[2] Ordnance Survey Library, box TL 23, item *Cassini co-ords one inch qtr in. half in.*

Horizontal neat lines in order:-

Sheet 1 north	2,263,727 feet north
Sheet 2 north	1,207,727 feet north
Sheet 1 south	890,927 feet north
Sheet 3 north	151,727 feet north
Sheet 2 south	165,073 feet south
Sheet 3 south	1,221,073 feet south

The Delamere origin is at the point marked on this series by the spot height 575, nine miles E by N from Chester. In accordance with the above limits, the co-ordinates of the sides of the twenty mile square containing Delamere, M10 on sheet 2, B10 on sheet 3, are:-

46,127 feet north

56,490 feet west   49,110 feet east

59,473 feet south

Can these be right, or are we already following the White Rabbit down the rabbit-hole?; let me apply some elucidation. The lines forming the two inch, eight scale mile, squares on the quarter-inch Third Edition maps coincide with every fourth line of the Popular Edition reference grid, and the ten-mile map squares might be expected to coincide with every tenth. But the north-south lines are displaced by one mile, falling centrally down Popular Edition squares; when it is realised that this brings the western neat line of the ten-mile series just 0.17 inch (4.4mm) outside Sròn an Dùin, the western point of Mingulay, and the eastern neat line 0.16 inch (4.0mm) outside Lowestoft Ness, it can be seen that this implies some accurate scheming.[3] The east-west lines, however, are a different matter; they are displaced southwards from such lines on the Popular Edition by 2,813 feet, or half a mile plus 173 feet, which latter increment is an insignificant 0.003 inch (0.08mm) on the map.

There appears no rational explanation for this shift in the horizontal lines of the ten-mile reference grid, which automatically brings an equal shift in the north and south limits of the map sheets. In fact, it brings a worsening in the positioning of these limits in that the clearance north of Seal Skerry, northern Orkney, is reduced from 0.18 to 0.13 inch (4.6 to 3.3mm), whilst that south of Gilstone Ledges, southern Scilly, is increased from 0.27 to 0.32 inch (6.8 to 8.1mm); a half mile shift *northward* in the limits would have evened these clearances to 0.23 and 0.22 inches (5.8 and 5.5mm) respectively! On the other hand, it may be no coincidence that with the strange recorded shift the latitude of the north-west corner of sheet 3 works out at 53° 32' 50".000 N, but there again seems no rational reason for this one of twenty-four geographical co-ordinates being made a round figure.

2. Two-sheet **Road Map**, 1932, and topographical map, 1937

The National Projection, already in embryo at the time the three-sheet series was published, was used in its original form for the construction of the succeeding pair of series. The National Projection was a Transverse Mercator Projection with its true origin at 49°N, 2°W, and as originally computed in imperial terms had a scale factor of 0.9996. This was the first Ordnance Survey projection to be manifested to the purchasing public by the incorporation of a numbered co-ordinate grid, initially appearing on the one-inch Fifth Edition in 1931. Although then known as the national grid, I use the full title National Yard Grid throughout

---

[3] This is the technical term, and no deviousness is implied.

to distinguish it clearly from its metric successor. As may be presumed, the unit of this grid was the yard, and the true origin was given the false co-ordinates 1,000,000 yards east, 1,000,000 yards north. The sheet line co-ordinates were:-

West neat line (1937)	620,000 yards east
West neat line (1932)	680,000 yards east
East neat line	1,300,000 yards east
North neat line sheet 1	2,180,000 yards north
Joining line sheets 1,2	1,650,000 yards north
South neat line sheet 2	1,120,000 yards north

The reader will hardly need me to point the difference between this orderly array of round figures, and the disorderly array which had resulted from the true origin of the projection being at an eccentric point within a grid square (to apply the term anachronistically to an earlier equivalent). I would add, however, that the earlier practice was in line with that used on the large scale, six-inch and twentyfive-inch, projections.

On the new projection the meridian of Delamere slopes at an angle of $0° 33'$, the convergence of the meridians, to the central meridian of $2°W$ and all north-south grid lines. As the topography from the former three-sheet series was transferred, presumably by manipulation of photography, on to the new projection it can clearly be seen, particularly in longer names such as Inverness-shire, Yorkshire and Irish Sea, that the "horizontal" work slopes downward to the right at the same angle; this is even more evident on the succeeding series with its closer spacing of grid lines. Being re-lettered, this problem did not afflict the **Road Map**.

3. 1:625,000 Ministry of Town and Country Planning series, 1942, and 1:1,250,000 sheet, 1947

Long before the appearance of the next ten-mile series, and indeed well before the Davidson Committee had completed its deliberations, the Survey had decided that it had no option but to adopt the international metre as the unit for its retriangulation. This necessitated recalculation of the National Projection on the same unit, making it very easy to fall in with the later Davidson recommendations that a national grid should be introduced with that unit.[4] The geodetic standards of the retriangulation required that a very accurate conversion factor was determined for feet of Bar O_1 to international metres, but as the time was still that of logarithmic computation this factor was defined by the absolute value of an eight-figure logarithm. In natural terms it is equivalent to 1 foot = 0.304,800,749,1 metres. The scale factor of this metricated National Projection was also defined by an eight-figure logarithm, equivalent to a natural value of 0.999,601,271,7.

The redefined National Grid now introduced on the modified projection, in addition to having its unit as the metre, had the false co-ordinates of the true origin amended to 400,000 metres east, -100,000 metres north. The formulae for conversion between National Yard Grid co-ordinates e,n and (metric) National Grid co-ordinates E,N are:-

$$E = 0.914,403,41\ e\ -\ 514,403.41$$
$$N = 0.914,403,41\ n\ -\ 1,014,403.41$$

$$e = 1.093,609,22\ E\ +\ 562,556.31$$
$$n = 1.093,609,22\ N\ +\ 1,109,360.92$$

---

[4] Viscount Davidson, *Final report of the departmental committee on the Ordnance Survey*, London: HMSO, 1938.

Applying these formulae to the limits of the 1937 two-sheet series above, we obtain:-

West neat line	52,526.70 metres east
East neat line	674,321.02 metres east
North neat line sheet 1	978,996.02 metres north
Joining line sheets 1,2	494,362.22 metres north
South neat line sheet 2	9,728.41 metres north

These when rounded to the nearest metre are seen to be the same as the neat line co-ordinates stated on the 1:625,000 and 1:1,250,000 maps, which were direct conversions from the 1937 series. The figures also demonstrate how the mainland of Great Britain falls neatly between the northings of 0 and 1,000,000 on the revised grid, one of the facts which the Ordnance Survey used to promote this metric intrusion into the imperial atmosphere of the time.

4. 1955 and subsequent maps

These series were constructed from the start on the National Projection with National Grid as currently in use. The defining elements of these have been recorded above, but for convenience I tabulate them here in the standard form:-

Projection	Transverse Mercator
Spheroid	Airy*
Unit	International Metre*
True Origin	49°N, 2°W
False Co-ordinates	400,000 m. E, -100,000 m. N
Scale Factor	0.999,601,271,7

* 1 foot of Bar O_1 = 0.304,800,749,1 metre

The sheet line co-ordinates are:-

West neat line	50,000 metres east
East neat line	670,000 metres east
North neat line sheet 1	990,000 metres north
Joining line sheets 1,2	500,000 metres north
South neat line sheet 2	10,000 metres north

3. The mixed parentage series

Twelve-sheet series, 1903, later eight-sheet

This series was directly reduced from the then current series of quarter-inch maps, themselves derived from one-inch maps. The problem was that these maps of England & Wales were on a different projection from those of Scotland, and the two projections could in no way be accurately fitted together at the border, or anywhere else; whilst east-west distances were broadly the same on the two projections, the curvature of the projected parallels of latitude was different.

The one-inch and quarter-inch maps of England & Wales were on Cassini's Projection on the origin of Delamere, as specified in section 2.1 above. The corresponding maps of Scotland at this time were constructed on Bonne's Projection on the origin of 57° 30' N, 4° W, with the foot as unit. The radius of projection of the initial parallel of latitude, 57° 30' N, was 13,361,612.2 feet, and the position of the origin was in the vicinity of Croy, some eight miles E by N from Inverness and on original Scotland one-inch sheet 84; the co-ordinates on Bonne's Projection of the neat lines of this map were:-

41,957 feet north

44,349 feet west 82,371 feet east

53,083 feet south

In another place[5] I distinguished as (a) and (b) the two intertwined problems I found in looking into the projection of a particular map series. I now define these items in more general terms, and with a marked lack of humility I christen them:-

Adams's Components

In the study of the projections of maps, as distinct from the study of map projections, two separate components are sometimes present; they should not then be confused, though their joint presence may confuse the issue. Such dichotomy is largely, but not entirely, restricted to maps from the era of national surveys, and on topographic scales. I label these components:-

(A) The projection upon which the sheet lines, graduation and graticule (where present) of the map or series are constructed; and

(B) The method used in laying down the topographical or other detail upon that framework.

Considering component (A) for the series under review, the sheet lines for the whole series were set out on the projection of the English larger scale maps; on this projection the central meridian of the Scottish projection was a very shallow curve, running at an angle to the north-south sheet lines varying from 1° 04' to 1° 09'. Turning to component (B), the quarter-inch maps of England & Wales were correctly reduced in on their own projection, but how were those on the discordant projection laid down? We are enabled to examine this question by the hindsight presented by the subsequent re-publication of the one-inch and quarter-inch maps of Scotland on Cassini's Projection on Delamere. From these we can trace where the detail on the ten-mile maps ought to be, and compare this with where it was actually drawn.

But first we need to obtain the co-ordinates of the sheet lines of the ten-mile series. I have not been able to find any extant documentation giving these figures, neither at the Ordnance Survey Office nor at the Public Record Office, so we have to resort to the maps themselves, remembering after the revelations in section 2.1 above that no certain values can be obtained therefrom. The sheet lines running east-west across England can be visually identified quite precisely with lines of the Popular Edition reference grid, and it is reasonable to suppose that these do coincide. But the central north-south join does not coincide with such lines, and tests show it to be running about one mile east of the meridian of Delamere; whilst it is only possible to estimate the position to some 250 scale feet, there is no reason to depart from the guess of exactly one mile east. Consequently, we arrive at the *estimated* co-ordinates of the sheet lines of the original twelve-sheet series as:-

Common west neat line	1,050,720 feet west
Central north-south joining line	5,280 feet east
Common east neat line	1,061,280 feet east
North neat line sheet 1,2	2,826,220 feet north

---

[5] 'The projection of the original one-inch map of Ireland (and of Scotland)', page 22.

Joining line sheets 1,2/3,4	2,139,820 feet north
Joining line sheets 3,4/5,6	1,453,420 feet north
Joining line sheets 5,6/7,8	767,020 feet north
Joining line sheets 7,8/9,10	80,620 feet north
Joining line sheets 9,10/11,12	632,180 feet south
South neat line sheet 11,12	1,344,980 feet south

The estimated co-ordinates of those sheet lines which were altered for the subsequent eight-sheet versions (which running in the sea, away from confirmatory land features, must be regarded as more suspect than those above) were:-

Sheet 1.2. (later 1) west neat line	401,280 feet west
Sheet 1.2. (later 1) east neat line	549,120 feet east
Sheet 3.4. (later 2) east neat line	216,480 feet east
Sheet 5.6. (later 3) east neat line	269,280 feet east
Sheet 7.8. (later 4) west neat line	546,480 feet west
Sheet 7.8. (later 4) east neat line	667,920 feet east

Returning to component (B) for the area of Scotland, I have traced the above co-ordinate lines across the one-inch Popular Edition maps of that country; from these it can be seen that, within the limits of accuracy of the ten-mile scale, the greater part of Scotland is laid down in its true longitude position, but is mainly placed nearly half a mile too far north, with slight variations attributable to the Bonne's Projection. The main shift is, however, much greater than can be accounted for by the different projection, whilst in the vicinity of the Anglo-Scottish border what should be straight co-ordinate lines are found to take on snake-like forms. These discrepancies cannot be fully explained, as no evidence has been found to show exactly how the reduced Scottish quarter-inch sheets were laid down; but although it is a commonplace that independently drawn maps of adjoining areas will not join precisely, it seems to me that the Scottish maps could have been positioned more accurately, and if so would have needed less fudging at the border.

4. The original (One-inch Index) series

Before considering the projection, or indeed any other aspect, of the very first Ordnance Survey map on the scale of ten miles to one inch, the reader should attempt to put him/herself into the state of knowledge of the shape of England and Wales in 1817; it is not particularly easy. I do not detract at all from the better efforts of the county surveyors, but there was no totally accurate knowledge of the disposition of any given area of the country until the relevant one-inch sheet was published. Nor was there any framework, beyond the area currently reached by the Principal Triangulation, upon which a small-scale map could be based. Thus one early map showing the progress of the survey[6] has some northern coasts badly displaced, and although north-east Norfolk is only two miles too far west this brings it wholly within a standard sized Old Series sheet 68; no doubt a similar map was used for scheming the main body of the sheet lines, with the result that sheet 68 is the only known entity to consist of six quarters.

It also needs to be explained that no knowledge survives of the actual method of construction of many of the early maps, and it is not generally possible to determine the

---

[6] PRO MPHH 1/239/1.

projection of such a map by inspection; the difference between the various projections used for constructing maps in the scale range we are concerned with is often less than the distortion which may occur in the paper during the printing process (or in fact in the drawing process in those days). It may however be possible to say from examination of the graticule, for example, which projections were not used, or it may be possible to make deductions from other items present on the map such as the sheet line pattern on an index sheet.

My main regret in compiling this appendix is that I have had to tackle this section without benefit of a detailed study of the projections, or lack of them, of the Old Series one-inch maps themselves, but those who have attempted to look into this matter, myself included, have been daunted by the sheer magnitude of the investigations which would be required. Nevertheless, an adequate assessment of the index sheets can still be made.

1. South sheet

Remembering the deprecating way in which this map was referred to as merely an index,[7] etc, in early literature on the subject, it comes as no great surprise when the first thing to be discovered about its construction is that the "rectangle" is not rectangular. In fact there are several features which lead me to suggest that the initial engraving was entrusted to an apprentice. Notwithstanding this, it would appear that the map was intended to be drawn on an accurate projection for, whilst the south-west corner was 3mm higher than the south-eastern, the measurements from each corner to equivalent latitude degree markings were the same on both sides. All degrees of both latitude and longitude except one were marked by prick-holes in the copper plate, which show as dots on printed copies, and all but three of these had ticks added outside the neat line; of varying lengths on the initial state and altered to a fairly uniform half-inch (13mm) on the next state, they were nearly all engraved at the wrong slope, usually shallower than that of the meridian or parallel they were supposed to be lying along.

Concentrating first on the initial state of this sheet, almost certainly dated 1817,[8] it could well be thought that the one-inch sheets indexed on this state and the topography contained within them, mainly south of the line of the River Kennet plus the county of Essex, were laid down on the same projection as the border (with the 3mm mis-plotting being somehow "accommodated"). There being no graticule on the copper, there was presumably one on a paper drawing upon which the one-inch limits were plotted before being transferred to the plate for engraving. This very first known proof is on a firm piece of paper showing little evidence of distortion other than direct paper shrinkage, and it is clear that the one-inch sheet rectangles do not come out as quite rectangular on the Index, and that the longer horizontal lines across groups of sheets on the same meridian exhibit very slight upward curvature thereon. These effects imply that the Index was on a different projection from that of the one-inch maps, with two equally unlikely possibilities; that the Index was on a conical-type projection with its origin in the south of France, or that there was something strange and hitherto unsuspected in the projections of the one-inch maps. But examination of the relative disposition of the western, central and eastern areas indicates a quite different projection again.

---

[7] PRO T1/9335c.

[8] An even earlier unfinished printing was discovered later, in 1995, in Lincolnshire Archives (reference Yarb. 4/29/7). Brian travelled to Lincoln to inspect it; his detailed report follows this paper.

INDEX MAPS

Projection: Transverse Mercator

Main sheet lines only are depicted; insets and extrusions are not indicated. The only overlaps are in the Great Britain three sheet series.

Origins of projections and grids

▲ Trigonometrical station
● Graticule intersection
⊙ False origin

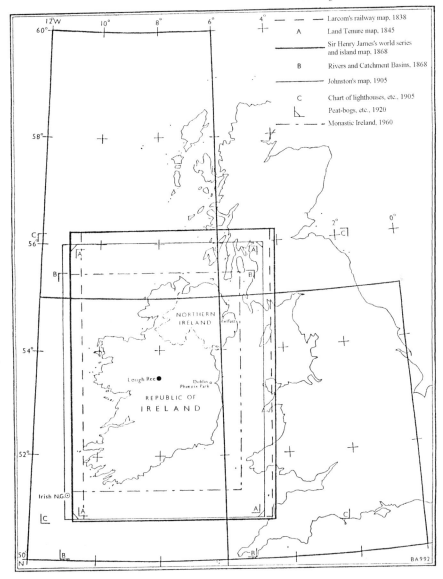

Larcom's railway map, 1838

A   Land Tenure map, 1845

Sir Henry James's world series and island map, 1868

B   Rivers and Catchment Basins, 1868

Johnston's map, 1905

C   Chart of lighthouses, etc., 1905

Peat-bogs, etc., 1920

Monastic Ireland, 1960

NORTHERN IRELAND

Belfast

Lough Ree

Dublin
Phoenix Park

REPUBLIC OF
IRELAND

Irish N.G.

BA992

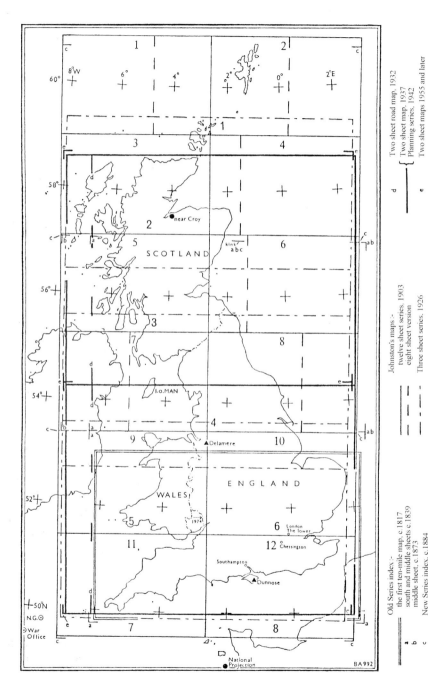

Old Series index :-
a ═══ the first ten-mile map, c.1817
b ─── south and middle sheets c.1839
c ─ ─ middle sheet, c.1873

New Series index, c.1884

Johnston's maps :-
─── twelve sheet series, 1903
─ ─ ─ eight sheet version
─ ─ Three sheet series, 1926

Two sheet road map, 1932
d { Two sheet map, 1937
{ Planning series, 1942
e Two sheet maps 1955 and later

SHEET NUMBERS are only shown for the twelve sheet and renumbered eight sheet series; three and two sheet series numbers (from 1884) run from north to south

BA 992

An attempt to trace the course of a latter-day straight line, a National Grid northing, across the face of the map produces some marked zig-zags; this is not unduly surprising as central southern England was surveyed and mapped after Essex and Kent to the east and Devon and Cornwall to the west, and had to be inserted between those areas. But a similar operation with a parallel of latitude makes it clear that its curvature is greater than would be the case if the Index were on Cassini's Projection, and implies a conical-type projection, possibly Bonne's, with its origin in the centre of England or of Great Britain. Returning to the degree ticks on the neat line, it is seen that the south-west corner lay on the meridian of 6°W and the south-east on that of 2°E so that the central meridian of this first ten-mile map was 2°W, giving shades of *déjà vu* to the National Projection of over a century later. From the 1820 state of the Index it is seen that the two northern corners were precisely on the parallel of 53°N, and from these various figures I calculate that the latitude of the mid-point of the south neat line would be latitude 49° 54' 35" N, if the Index were on Cassini's Projection, but 49° 55' 00" if it were on a Bonne's Projection with its origin on 53°N; the round figure seems the more in keeping with the exact degree values at the corners as noted above but, since a shift of the origin to 55° 30' N would only drop the south neat line by 0.1mm, nothing positive about the origin is suggested.

On the next few states of the south sheet new one-inch sheets were added piecemeal but presumably on the original projection. However, when the plain neat line came to be deleted and a piano key border added farther out, c.1839-41, something most peculiar also happened; an attempt seems to have been made to change the projection in mid-stream, not of course that such a thing is possible, at least for the material already *in situ*. On the early states of this sheet carrying the new border, a score-mark of a revised meridian can be seen just over 1mm to the west of the 2°E meridian in the south-east corner; the new border itself carried no graduation whatsoever. I should also add that the neat line of the piano key border in the south-west corner was engraved just over 3mm farther below the old neat line than it was in the south-east corner, thus slightly over-correcting the lack of rectangularity in the original.

From further checks on sheet lines and the paths of National Grid lines and parallels of latitude across the sheet, it does seem that the material added with and later than the new border was on a different projection to the earlier material, very possibly Cassini's. Once again, however, discontinuities occur across Lincolnshire, surveyed and mapped in advance of the adjoining areas. Examining now the New Series index version of this map, it would appear that all the sheet lines of this series were engraved as though the entire basic map was on Cassini's Projection, even though the southern counties had almost certainly been drawn on a different projection. Added to the distortion which produced the zig-zag effects already referred to, this meant that the sheet lines, particularly in the south, were badly placed in relation to the topography they cut through. The neat lines of the new border which was engraved on the New Series version were realigned to be parallel and perpendicular to the meridian of Delamere, the central meridian of this series as mentioned in section 2.1.

2. Middle and north sheets

There being no graduation on these sheets from the start, all that is possible here is to examine the sheet patterns, bearing in mind that the situation was akin to that already examined in section 3 above – the one-inch sheets of England were on Cassini's Projection whilst those of Scotland at the time were on Bonne's Projection. As the sheet lines in both countries appear as true rectangles, it seems that we again have the case that the two areas were drawn on the two different respective projections, both appearing together on the middle sheet; in this instance, however, the two countries seem to have been accurately laid down relative to each other. As with the south sheet (New Series), and with the subsequent twelve-sheet series, the neat lines of the middle and north sheets of the New Series index were parallel and perpendicular to the meridian of Delamere.

5. The Mainly Mythical Projection

G B Airy, later Sir George Airy, Astronomer Royal 1835-81, made a number of notable contributions directly and indirectly to the scientific work of the Ordnance Survey. His name appears very early in this appendix as having calculated the figure of the Earth, or spheroid, which has been used for the vast majority of all OS surveys and mapping; it was first published on 17 August 1830 when Airy was still Plumian Professor of Astronomy at Cambridge, but is thought to have been communicated to the Survey a little before that.

During the summer of 1861 Airy's active mind thought up an idea for an improved map projection[9] and this was published by him with the name of a Projection by Balance of Errors;[10] amended slightly by James and Clarke it is known today as Airy's Projection.[11] Plainly described by Airy as for maps of a very large extent of the Earth's surface, and recommended by more than one authority as deserving of more attention, it has in fact received very little use, no doubt due to the rather complex computations necessary to construct it.

However, the then Major C F Close stated[12] that it had been used for the Ordnance Survey ten-mile map of England (note specifically England) and that statement has been repeated by many authors right to the present day; I have not been able to trace any prior such reference. But the only ten-mile map of England published between 1861 and 1901 was that of **Rivers and their Catchment Basins** 1868 and whilst, as remarked in 4 above, it is not possible to determine absolutely the projection of such a map by inspection in the absence of any details of its construction, it is inconceivable that this map was on Airy's Projection. For Sir Henry James was personally involved in the conception of the 1868 map, and Sir Henry knew very well that Airy's Projection was only intended for very small scale maps since he had invented a similar projection himself and he and Airy had corresponded about the two projections, each defending his own as the better.[13] Furthermore, the form of Close's reference in 1901 seems to imply a regular map rather than a particular thematic issue.

Nor was Close's remark apposite to a map then in the course of preparation, even had we not known that the 1903 map was on a different projection for England (and another for Scotland), so to what then was Close referring, for we know he would not have made such a statement lightly. It would seem that he must have been repeating something garnered during his first spell of duty at the OS Office in 1897-98. I now move forward to 1934 when Brigadier H St J L Winterbotham as Director General wrote:[14]

9   Cambridge University Library, reference RGO 6/435, 6/475.
10 Airy, G B, 'Explanation of a projection by balance of errors for maps applying to a very large extent of the Earth's surface; and comparison of this projection with other projections', *Philosophical Magazine*, 22 (1861), 409-421.
11 James, Colonel Sir Henry, Clarke, Captain A R, 'On projections for maps applying to a very large extent of the Earth's surface', *Philosophical Magazine*, 23 (1862), 306-312.
12 Close, Major C F, *A sketch of the subject of map projections*, London: HMSO, 1901.
13 Cambridge University Library, reference RGO 6/472.
14 PRO OS 1/144.

It has always been, in my time, a tradition that the ten mile map of Great Britain (sic) was originally upon Airy's projection. At the end of my time as OTT here I began to be extraordinarily doubtful,

and requested the Research Officer, Harold L P Jolly, to *"substantiate"*. Jolly submitted a lengthy paper effectively coming to similar conclusions to those above, but also quoting a minute by Major Alan Wolff in 1919:

There are very old drawings in the MS store of an index of old series 1-inch maps on 10 mile scale which is quite different from any of the existing copper plates and it seems very probable that this may be on Airy's projection but there is nothing on the drawings to indicate the central point or the circular area included in the projection.

Jolly dismissed Wolff's suggestion, but without adequate reason apparent from this distance of time, and the "very old drawings in the MS store" may well be the source for Close's 1901 statement; no other possible source is currently evident.

However, as to the basic "tradition", Winterbotham accepted that "The myth of the Airy Projection is now laid", but also handed down the decision that Jolly's memorandum "washes too much soiled linen for publication." So it is only now that it appears in print for the first time that no Ordnance Survey map on the ten-mile scale was ever published on Airy's Projection. I have to add that particular thanks are due to Dr Richard Oliver for first bringing to light the PRO document detailing this matter.

6. The Irish ten-mile maps

1. Larcom's map, 1838

Professor John Andrews[15] has not been able to find any record of the actual method of construction of this map, and the chances of anyone else so doing must be regarded as microscopic; he does however suggest, no doubt rightly, that the mathematical basis would have been set out by Captain Joseph Portlock. The only positive thing that can be said about the projection is that the central meridian was 7° 50' W; otherwise, even less can be learned from this map than were a complete graticule present. Speculation as to the projection used can only be based on what was fashionable at the time, and even the projections used for related scales were not particularly relevant at this date; its place is therefore in the realm of day-dreams, and not in this appendix.

2. Sir Henry James's world series

The projection used by Sir Henry for this ambitious project had the best contemporary documentation of any of the early projections employed by the Ordnance Survey, but the material has to be read with circumspection and translated into modern terms, the main trap being that two slightly different projections were involved, whilst a third had also been suggested by Sir Henry for world-wide use for larger scale maps. The latter, which he distinguished as "the purely tangential projection",[16] was that known today as the Simple Polyconic Projection, and from this James O'Farrell evolved an alternative form of construction to produce the Rectangular Tangential Projection, today's Rectangular Polyconic

---

[15] Personal discussion.
[16] James, Colonel Sir Henry, *On the rectangular tangential projection of the sphere and spheroid, with tables of the quantities requisite for the construction of maps on that projection; and also for a map of the world on the scale of ten miles to an inch, with diagrams and outline map*, Southampton: Ordnance Survey, 1868.

Projection; these names reflect the feature that the meridians and parallels cut at right angles, but the projection is not conformal.

It was first used for very small scale maps of whole continents and the world, but when Sir Henry James proposed its adoption for a systematised scheme of sheets to cover the world at the ten-mile scale it was again O'Farrell who calculated a modification, claimed to reduce the average distortion of projected distances in both principal directions to zero; this Modified Rectangular Polyconic Projection is not to be confused with one written up by G T McCaw in 1921. O'Farrell's published computations only permitted the construction of sheets conforming with the standard scheme of sheet lines; they were calculated on the Clarke 1858 spheroid, and the centre line of each sheet was the central meridian of its projection. Across each sheet the meridians of longitude were drawn straight, so that the convergence of a particular meridian on that sheet was constant, but on the adjoining sheet towards the Equator the angle would be smaller and on the next sheet away from the Equator it would be greater.

3. Sir Henry James's island map, 1868

This single sheet map of Ireland was not on a proper projection; it was formed by butting together the three Irish sheets of Sir Henry's world scheme, each *individually* on the projection described above, leaving the north-east corner to be filled in separately. Consequently there were changes in the directions of the meridians (except for the original central meridian of 9°W) across the sheet join along 55°N, producing slight dog-legs in those lines. The maximum kink occurred in the sheet edge meridian of 6°W, and although the change of direction was only 0° 09', this can be clearly seen with the aid of a straight-edge on those maps which retain a graticule. A new rectangular border was constructed around the island of Ireland with its west and east neat lines parallel to the meridian of 8°W, adopted in the previous decade as the central meridian for the one-inch series of Ireland (on a different projection, see below). A graduation was established in this border, presumably derived from its cuts across the graticule, and this was completed, without any change in the spacing of the ticks, round the initially blank area north and east of 55°N, 6°W; by joining the relevant ticks to the existing graticule on the rest of the map, a rather distorted framework was produced within which the Firth of Clyde, Sound of Jura and adjoining areas were drawn in skeleton form.

4. Johnston's map, 1905

This map was constructed on what had become the regular national projection for Ireland, originally adopted for the country's one-inch maps, almost certainly on the recommendation of Captain William Yolland. These were constructed on Bonne's Projection on the Airy spheroid, on the origin of 53° 30' N, 8° W, the unit of calculation being the foot; the radius of projection of the initial parallel of latitude, 53° 30' N, was 15,516,209.8 feet, and the position of the origin was just in the waters of Lough Ree on Ireland one-inch sheet 98. The co-ordinates on Bonne's Projection of the neat lines of the latter were:-

35,720 feet north

27,040 feet west   68,000 feet east

27,640 feet south

The above elements specify component (A) for the Ireland one-inch series, and on page 23 I explain the incorrect use of a component (B) for compiling that series from the six-inch maps. However, the maximum error resulting from this misuse was 0.004 inch (0.10mm) on the one-inch scale, which is almost invisible, and becomes totally invisible when reduced again to the ten-mile scale. Johnston's map may therefore be said without reservation to be drawn on the Bonne's Projection specified above.

5. 1:625,000 Monastic Ireland, 1960

This was the first ten-mile map of Ireland to be presented on the metric version of the scale, on the Transverse Mercator Projection of Ireland, and carrying the Irish National Grid; there was however no mention of the latter fact on the face of the map and numbering of the grid lines was limited to the 10km figures, repeating every 100km across the map. The attached pages of text contained brief notes on the lettering of the 100km squares (known as sub-zones) and on grid references; for the second edition the notes were longer but rather convoluted. The defining elements of the projection and grid as first introduced were:-

Projection	Transverse Mercator
Spheroid	Airy*
Unit	International Metre*
True Origin	53° 30' N, 8° W
False Co-ordinates	200,000 m. E, 250,000 m. N
Scale Factor	1

* 1 foot of Bar O_1 = 0.304,800,749,1 metre

The actual sheet line co-ordinates are:-

West neat line	10,000 metres east (numbered 1)
East neat line	370,000 metres east (numbered 7)
North neat line	470,000 metres north (numbered 7)
South neat line	10,000 metres north (numbered 1)

When the retriangulation of Great Britain was carried out, from 1935 onwards, it was decided early on to avoid disturbance to existing material as far as possible, and steps were taken to carry the scale and orientation of the old Principal Triangulation through to the new Primary Triangulation and new mapping based thereon. But after completion of the retriangulation of Ireland in 1964, and its adjustment in the following year, it was decided between the Irish surveys to alter the scale of their triangulation and mapping to bring distances determined therefrom into line with electronically measured true distances. This was effected by altering the scale factor of the projection and grid from unity to 1.000,035 with an equal and opposite alteration in the dimensions of the fundamental spheroid, in order to avoid recalculation of the projection tables, co-ordinates etc. Reduced by 0.999,965,001,2 it was christened the Airy (Modified) spheroid.

Subsequently, it appears that Northern Ireland had second thoughts on the latter part of these alterations, and by 1983 the six counties had reverted to the use of the true Airy spheroid.[17] As the new scale factor could not be interfered with, the projection was concurrently re-christened the "Modified Transverse Mercator Projection"; this appears to me to be nonsense, there is no modification in the projection at all. I do not think either of the Irish surveys' solutions was right, and they were surely unnecessary today when software is available which can instantaneously calculate distances on the Earth, on in terms of other

---

[17] *The Irish Grid*, OSNI leaflet 1, July 1983.

national or international datums. I therefore leave the listing of the projection elements to "as first introduced" above.

7. Military grids

With the exception of the National Grid, which in this context may be regarded as a dual-purpose grid, all the military grids which appear on OS ten-mile maps are on different projections or different origins from the maps on which they are superimposed. An attempt to explain this apparent contradiction to the uninitiated has been made by me.[18]

1. War Office Cassini Grid (military title: English Grid)

Projection	Cassini's
Spheroid	Airy
Unit	Metre (1 foot of Bar O_1 =0.304,799,73 m.)
True Origin	Dunnose (50° 37' 03".748 N, 1° 11' 50".136 W)
False Co-ordinates	500,000 m. E, 100,000 m. N
Scale Factor	1
Grid colour	Purple

Co-ordinates on this grid are known as War Office False Origin (familiarly Woffo) co-ordinates.

2. War Office Irish Grid (military title: Irish Grid)

Projection	Cassini's
Spheroid	Airy
Unit	Metre [19]
True Origin	53° 30' N, 8° W
False Co-ordinates	199,990 m. E, 249,975 m. N
Scale Factor	1
Grid colour	Red

The false co-ordinates of the origin were originally 200,000 m. E, 250,000 m. N, but were altered after a War Office adjustment of trigonometrical station co-ordinates.

3. Continental grids

All of these grids of which portions fall in the sea on ten-mile maps of Great Britain are on the Lambert Conformal Conical Projection and have the metre as their unit; they have a variety of spheroids and scale factors and a full listing of their elements does not seem relevant here. The grids concerned and their colours are:-

French Lambert Zone I	Red
Nord de Guerre Zone	Blue
Northern European Zone III	Blue

Appendix 1 to Roger Hellyer, The 'ten-mile' maps of the Ordnance Surveys,
London: Charles Close Society, 1992.

---

[18] In '198 years and 153 meridians, 152 defunct', see page 53.
[19] Conversion factor as in 7.1, the Benoit and Chaney relationship.

The earliest known[1] Ordnance Survey ten-mile map
First report, from visit by Brian Adams, 3 April 1995

Lincolnshire Archives (new building 1991)

St Rumbold Street, Lincoln LN2 5AB
NG Ref: SK 979713
Phone: Search Room   01522 525 158
        General enquiries   01522 526 204
Hours: Mon 1.30-7.15; Tue-Fri 9-5; Sat 9-4.
For Reader's Ticket: 2 ID photos
                       Proof of identity (not asked of me)
                       Proof for concessions (e.g. OAP)
One-day visit fee: £1.20
Visit should be arranged, and item ordered, in advance.
Restrictions: fairly standard, pencils only,
                       no folders (except transparent ones),
                       no books (but Hellyer + Adams was OK),
                       no food, not even in entrance hall.
Several sizes of lockers, requiring 50p coin.
Photo-copying not available in house, and takes two weeks!
(no full sheet copier, normal A3 only)

The 'new' index

is in a roll containing items Yarb. 4/29/5-7. Items 4/29/5,6 comprise the Old Series Parts V, VI, with 4/29/7 being the Index; this implied confirmation that only Parts I to VI were obtained at that time is reinforced by the colouring (see below). The index is in outstanding condition, except for the edges of the paper.

Sizes in mm :-

		1013				959				892	
Paper	677		677	Plate	654		660	Neat line	550		550
		1019				964				895	

indicating a further lack of rectangularity; another visit with millimetric grid is vital.

Watermark :-

1809 (in corner), H S (in centre) on north side, possibly with a badge of some sort; to view it, twist round the illuminated magnifier in Carrel no. 1 (next to search room door).

Rectangle and Graduation :-

In contrast with the St John's copy (S-P1), the rectangle is complete, and those degree ticks which are present are all half an inch long (as on the subsequent copies), showing that they were on the copper from the start. A scratch outside the neat line south of Portland, which can be seen on S-P4 but not S-P1, is also present, all this indicating that

---

[1] As at March 1995.

the inking up of the copper for the St John's pull was even more limited than would be imagined.

The hope that ticks or prick-holes sight be found on or near the north border was not realised, except that the central meridian of 2°W appears fairly faintly from a prick-hole $^5/_{16}$ inch outside the north neat line right to the plate edge. Only this and its southern counterpart are at right angles to the neat line; all others are at the slopes referred to in Hellyer + Adams, page 180, opening paragraph of section 1.[2]

Colouring :-

The sheet is very nicely coloured by counties to the one-inch sheet limits, but the areas of sheets 19, 14, 6, 3 (Old Series parts VII, VIII), plus the part of sheet 7 shown, are all left untinted. As indicated by Richard Oliver and above, these are the areas not included in Yarb. 4/29.

Content :-

I did not have time to examine this in its entirety, but so far as I could judge, *excluding names*, the content of the Lincoln sheet was virtually identical with the St John's sheet, except that some ancient trackways in Wiltshire were not shown, and most roads in the south-east were only lightly engraved to be re-entered later. In particular, the detail shown in the east of sheet 7 was the same on both specimens.

The names of towns, i.e. those shown in capitals, were virtually complete, and the positional crosses of the lower case villages appeared to be all *in situ*. The names of the latter were mainly complete in Cornwall, Devon, Hampshire and Sussex, but within these counties some three dozen remained to be inserted, for no particular apparent reason. Whilst locational horizontals had been lightly engraved, the lower case names were conspicuous by their absence throughout Essex, Kent, Surrey and Berkshire, as well as in parts of the remaining counties of Somerset, Dorset and Wiltshire.

Conclusion

To sum up, it is clear that the Lincoln sheet was a pull taken from the partially engraved copper, presumably taken at that time to accompany Old Series sheets then being supplied, and that its overall content was to be that which appears on the St John's state, S-P1 (apart from the defective surround on the latter). With that proviso, S-P1 is therefore confirmed as the original intended full state of the ten-mile scale, one-inch series index, and the probability appears to be that it was put in hand at the time of the resumption of public sales of that series.

Brian Adams
Parsons Green
5 April 1995

Unpublished report to Roger Hellyer

---

[2] Reproduced on page 11 of this volume.

The projection of the original one-inch map of Ireland (and of Scotland)

Doubts have been raised from time to time concerning the projection used for the one-inch map of Ireland, that is the original 205-sheet series dating from 1855 and published in a number of forms, but now as a whole destined to remain the only all-Ireland one-inch series ever published. The doubts about this series, generally believed to be on Bonne's Projection, have been summarised by Professor John Andrews on pages 231-3 of *A paper landscape*,[1] but to the best of my knowledge no such doubts have been expressed on the original one-inch map of Scotland, first published concurrently with the Irish series and also held to be on Bonne's Projection; the first three editions of the Scottish one-inch were all on the same projection and sheet lines.

On looking into this matter more closely I find there are two problems intertwined:

a) the projection used for the sheet line system and the associated latitude and longitude graduations; and

b) the positioning of the topographical detail on the sheets.

On question (a) the fact that the data for the projections and their origins, in particular the projected radii of the parallels of latitude of the origins, were firmly stated in OS publications leaves little space for reasonable doubt. But in 1905 the former OS Director General, Colonel Duncan Johnston, told a meeting of the British Association held in South Africa that 'the 6-inch and larger scales, and also the 1-inch maps of England and Ireland, are on what is known as the rectangular tangential projection', then describing what is certainly not the rectangular tangential projection but, although in rather vague terms, may be presumed to be Cassini's Projection; he continued, 'The 1-inch of Scotland is on a modified equal-area projection known as Flamstead's Modified'.[2]

In considering these statements it is relevant to remember that Colonel Johnston's particular sphere of interest was in reproduction methods and not geodesy, that he had been responsible for issuing the second edition of *Methods and processes*,[3] and that he had probably composed his paper aboard the Union-Castle liner without some requisite reference works; further that *Methods and processes* described the projection used for the one-inch map of England and the six-inch maps of counties without naming it, that this projection (Cassini's) was always referred to by the OS as 'projection by rectangular spheroidal co-ordinates', and that the book states that the one-inch of Scotland is on a different projection, Flamstead Modified (i.e. Bonne's, see note), but does not mention the one-inch of Ireland. I therefore conclude that Johnston inserted the name of the English projection from memory, mixed up his rectangular projections in so doing, and added mention of Ireland but wrongly attached it to England instead of Scotland; and consequently I discount these references entirely.

---

[1] Andrews, J H, *A paper landscape: the Ordnance Survey in nineteenth-century Ireland*, Oxford University Press, 1975; see also the same author's contributions to W A Seymour (ed.), *A History of the Ordnance Survey*, Folkestone: Dawson, 1980.

[2] Johnston, Col Duncan, 'A brief description of the Ordnance Survey, and some notes on the advantages of a topographical survey of South Africa', *Scottish Geographical Magazine*, 22, 1 (1906).

[3] James, Sir Henry (ed.), *Account of the methods and processes ... of the Ordnance Survey ...* , London: HMSO, 1875.

However, for added certainty I have calculated the positions of two of the outermost points of the Irish one-inch sheet line system on the different projections, and compared them with the actual sheet graduations. The two points are the north-east corner of sheet 21 and the south-west corner of sheet 197, and at these extreme points there is sufficient north-south separation in the construction of the various projections to prove that the system of sheet lines is not drawn on Cassini's Projection nor on the rectangular tangential projection, and that it can be and virtually certainly is on Bonne's Projection.

Turning to problem (b), the practical cartographer from the days of pen, ink and all sorts of paper, would probably admit that the actual drawing in of detail could involve some *ad hoc* methods, with the hopeful end that errors of position remained less than the possible paper distortion; it might therefore seem somewhat pointless to pursue this matter. However, as all the detail (other than revision material) on the series under review was reduced in from county series six-inch maps, a single basic method should have been used to effect the transfer from county Cassini projections to national Bonne projections before the matter was engraved on the one-inch scale. It may be useful here to remind readers that all the six-inch sheet lines were Cassini co-ordinate lines of their respective county projections, and all the old Irish and Scottish one-inch sheet lines were co-ordinate lines of the national Bonne's Projections; also that a transferred six-inch rectangle would theoretically not be quite rectangular on the one-inch map, but that the actual difference would be very far indeed from a visible quantity.

The correct procedure then should have been to calculate the geographical positions (latitudes and longitudes) of the corners of each six-inch sheet from their Cassini co-ordinates referred to the county origin, then to convert these positions to rectangular co-ordinates on Bonne's Projection referred to the national origin, whence the corners of the six-inch sheet would be measured out and marked in on the new one-inch plate, and the detail from that six-inch sheet reduced in to the reduced rectangle so established. I am greatly indebted to two Dublin professors who supplied me with copies of some of their research material concerning this and related matters, and both of whom included descriptions of other methods supposedly used for laying down the six-inch sheet corners.[4] This revealing material was received after I commenced work on this paper, and I hope it will be of interest to members if I record my subsequent research step-by-step.

Professor John Andrews, now retired to Gwent, provided me with a minute written by the then Captain H St J L Winterbotham in 1914[5] which first confirms my conclusion re (a) above, then states unequivocally that the geographical positions of the six-inch corners in Scotland and Ireland were correctly calculated but were wrongly converted to *Cassini* co-ordinates on the relevant national origin, using the familiar Clarke's formulae. If this were so the work would be out of place relative to the one-inch sheet graduations by amounts increasing with diagonal departure from the national origin, and rising to a displacement of 0.10 inch too far north at Muckle Flugga. Such an error would be easy to spot, and could have been corrected for a detached island area, but I checked a number of widely spaced trigonometrical stations on the Scottish and Irish mainlands as well as in the Shetlands, and I found no evidence of any systematic error. I therefore have to discount Winterbotham's statement, except for his odd remark that the wrong method was still in use (in 1914), even though the last of the basic one-inch sheets had been completed over a quarter of a century

[4] *Latitudes & longitudes of the initial points of Ireland*, MS, Ordnance Survey of Northern Ireland, Belfast.
[5] Winterbotham, Capt. H St J L, minute dated 9 October 1914 in OSL 16769 (1914-15), National Archives of Ireland.

earlier. If he was referring to the reducing in of revision material on six-inch sheets, this opens up visions of adjoining detail being several hundredths adrift, and brings home my earlier references to ad hoc methods.

Professor Thomas Murphy had meanwhile sent me evidence of his investigations into the Irish six-inch maps, in which he had had the advantage of having seen the Ordnance Survey manuscript 'Calculation sheet book'.[6] Whilst this does not contain all the OS calculations, it does give the results of their calculations, and these were compared with the results of his own calculations. He found a systematic error in the north-south direction to be emerging between the figures he arrived at using the correct procedure I have stated above, and the figures in the OS book, and it was then suspected that the OS had used the following method: the geographical positions of the six-inch corners had been calculated from the county Cassini co-ordinates as though they had been rectangular co-ordinates on a Bonne's Projection with its origin coincident with the county origin. Re-calculation of a number of representative points by this method gave a concordance with the figures in the OS sheet book to the second place of decimals, thereby showing beyond reasonable doubt that this method had been adopted by the OS. In all these calculations the provisional positions of the origins and the county twists referred to on page 79, were used.

Although we know that Bonne's Projection was fashionable at the time of the preparation of the Irish one-inch map, the use of it in this way was strictly incorrect and it is difficult to imagine exactly how this came about. However the maximum error resulting from this misuse of the projection occurs at the outermost corner of the largest county, County Cork, and is only 0.004 inch or less than half a hundredth, which is usually regarded as an acceptable degree of accuracy in drawing on paper. Whether this possible discrepancy was calculated out in 1855 can only be speculated upon.

Armed with the news of this method of calculation for the Irish counties, I returned to my happy hunting ground in class OS 2, trigonometrical records, in the Public Record Office at Kew, in the hope of finding what method was used in Scotland. Having obtained the book containing the results for the largest county area, the unified counties of Argyll and Bute, I was pleased to find it also contained all the calculations as well as the results for another large county, Sutherland. The figures for these counties showed conclusively that the correct method and not the 'Irish' method had been used to calculate the geographical positions of the six-inch corners in Scotland, those for Argyll and Bute being quoted to and correct to three places of decimals. But what emerged from the complete Sutherland figures brought yet another surprise – the final calculations provided rectangular co-ordinates on the national Bonne's Projection in units, not of feet, but of fathoms! Whilst this was somewhat gratifying to a retired hydrographic officer, it was initially rather baffling in an OS context; however, due thought suggested that this apparent quirk would enable the measurements to be directly plotted on the one-inch scale, utilising some instrument(s) graduated in **scale** feet at six inches to a mile. It was Ian Mumford who reminded me of the scoring machine, an early form of co-ordinatograph described and illustrated in *Methods and processes*, in which the indications are that co-ordinates in feet could be directly read off on its brass scales when

---

[6] *Calculation sheet book for one inch map*, MS, Ordnance Survey Office, Dublin; Murphy, Thomas, 'The latitudes and longitudes of the six-inch sheet maps of Ireland', *Geophysical Bulletin*, 13 (1956), Dublin: Institute for Advanced Studies; Murphy, Thomas, 'Notes on the six inch and one inch sheet maps of Ireland and methods for deducing rectangular and geographical coordinates for points thereon', *Geophysical Bulletin*, 39 (1988), Dublin: Institute for Advanced Studies.

engraving new six-inch copper plates. So was yet another satisfactory solution arrived at, literally on the eve of the editor's deadline for this number of *Sheetlines*, and so for the time being I leave this subject to rest.

NOTE: *Methods and processes* says 'The projection for the one-inch map of Scotland differs from that of England ...[7] It is the same as that used for the map of France, and sometimes called Flamstead Modified', and follows this with a description of the projection which confirms that it is the one known to us as Bonne's. I have not come across any source for the name Flamstead Modified, but Flamsteed's Projection (this spelling is correct) is one of several names which have been used for the Sinusoidal Projection, which is in fact the particular equatorial case of Bonne's Projection.

Part I of 'An Irish miscellany'
Sheetlines 30, April 1991

---

[7] James, Sir Henry (ed.), *Account of the methods and processes ... of the Ordnance Survey ...* , London: HMSO, 1875.

General Roy's baseline

In Sheetlines 38 Lez Watson asked for information on the National Grid references for the end points of General Roy's baseline on Hounslow Heath and David Archer supplied an answer from his maps. This led to the following in the next issue ...

Brian Adams writes that he thinks that this subject is one where 'back to basics' really is the correct approach to calculating the accurate length of General Roy's baseline. We need to start with the positions of the base terminals taken from the old Principal Triangulation, which are:

King's Arbour: Lat 51° 28' 45".489 N, Long 0° 26' 55".454 W
Hampton Poor House:[1]   Lat 51° 25' 33".435 N, Long 0° 21' 50".975 W

It is then necessary to apply corrections of +0".015 latitude and -0".118 West longitude to bring these positions into sympathy with the retriangulation positions in the area. Conversion of the amended positions into National Grid co-ordinates gives the results:

King's Arbour: 507,715.15 metres E, 176,789.34 metres N
Hampton Poor House: 513,719.99 metres E, 170,985.21 metres N

These provide grid references of TQ 077768 and TQ 137710 respectively, but it should be noted that these apply to the true positions of the two terminals, which may not be exactly where drawn on a particular map! However, David Archer's readings from the map differ in one place only from the calculated references.

Calculation of the true distance (*not* the grid distance) between the above positions results in a figure of 8,353.48 metres, or 27,406.37 feet, which compares favourably with the length determined by Clarke in 1858 of 27,406.19 feet. This was one of the three lengths recorded on the tablets mounted on the gun barrels marking the base terminal stations in 1926. A transcription of the wording on the plate on the Hampton Poor House terminal in Roy Grove, Hampton, accompanies this note. It seems very probable that it was the bicentenary referred to on the tablets which prompted the addition of the baseline and its terminals to the Popular Edition sheet 114, on which they first appeared in a 1929 printing.

A copy made by Brian Adams of the plate on the cannon at Hampton Poor House.

Sheetlines 39, April 1994

---

[1] In some OS sources 'Poor House' is rendered as one word.

Hounslow Heath, Hampton and Heathrow
Roy re-visited and re-measured

The celebrations of the 250th anniversary of Military Survey have been previewed in recent numbers of *Sheetlines*. Billed as the main launch event was the re-measurement by modern methods of Major-General William Roy's historic Hounslow Heath baseline, itself the subject of several references in *Sheetlines* during 1994. The original measurement of this baseline had played a most important part in the development of ordnance military surveying, before that common operation began to divide into two separate streams and to acquire initial capitals.

Briefly, the base had initiated the English side of the triangulation to connect the observatories of Paris and Greenwich and subsequently became the starting line for the principal triangulation of the British Isles. Measured by three different methods in 1784 and again in 1791, by 1858 eleven different lengths were available for it, deriving from actual measurements, reductions to mean sea level, different definitions of the foot and triangulation adjustments, noting however that it was not one of the bases utilised by A R Clarke in his definitive adjustment of the Principal Triangulation published in 1858. The Retriangulation of Great Britain produced another calculated length, but it should be made clear that such a figure could in no way supplant an accurate direct measurement.

On 1 July 1997 a party from 19 Specialist Team RE, Royal School of Military Survey, Hermitage, Berks, assembled in Roy Grove, Hampton, to re-measure the baseline using GPS (Global Positioning System) equipment. Preliminary work had already been carried out, such as removing commemorative tablets, erecting protective tents and (off the record) doing a measurement. A small exhibition area of the work of Military Survey was erected on the grassed area at the head of Roy Grove. For those unfamiliar with it, this is a short cul-de-sac with the Hampton Poor House (south-eastern) base terminal at one side of its head.

First inspection of the army team was carried out by the red-jumpered members of the Tadpoles Nursery School which meets in Roy Grove and who were most cordially received by the soldiers. Next came the official party from Military Survey headquarters at Feltham, which included Sir Alan Muir Wood, representing the Royal Society and Colonel André di Martino, head of the Centre Géographique Inter Armées, in recognition of the original Anglo-French operation. After an introduction by the Director of Military Survey, Brigadier Philip Wildman, the operation was described by Major Alan Honey, in command of 19 STRE. The instruments were read and the distances calculated by Sergeant Andy Gray.

The GPS equipment provided a slope distance between the two terminals of 8,353.572 international metres, with an altitude difference of 9.388 metres, differing slightly from that obtained in 1791. From these figures was derived a mean sea level length for the baseline of 27,406.56 feet of O_1, referred to the WGS84 spheroid used by GPS. Whilst this compares very favourably with the equivalent 1791 Mudge/Clarke figure of 27,406.19, Sgt Gray also recalculated the latter using the WGS84 spheroid and the 1997 altitude difference and obtained a length of 27,406.61 feet of O_1, a most remarkable confirmation of the accuracy of Captain William Mudge's measuring procedures. No comparable length figure has hitherto been available for General Roy's measurement in 1784, but I have now calculated it to be 27,405.96 feet of O_1 which in 1997 terms would be 27,406.38, representing an apparent error of just 2.2 inches in 5.2 miles.

Not included in the official party, but probably as knowledgeable as any of them, were Dave Watt and myself, and after the proceedings at Roy Grove Dave very kindly chauffeured me to the north-western base terminal of King's Arbour. This lies between the northern perimeter road of Heathrow Airport and its police station and because of the traffic conditions Dave was unable to tarry there but just let me down. The cannon is at one side of a small area of bushes, but the most surprising event of the day in this area of continuous air and road traffic noise was to be preceded up the path by a large rabbit which left me in no doubt as to its resentment of my human presence by repeatedly turning and glaring at me, eventually disappearing into the bushes.

I found the cannon here to have been painted silver, checked the wording on the commemorative plate (one transcript having contained an error), and then moved on to the police station. Here on the south wall, is a further tablet erected at the instance of Major-General R C A Edge in 1968, when the cannon was in store and there was no above-ground evidence of the base terminal, and stating that the terminal was 109 yards to the south. I then made my way home, highly satisfied with this celebration of the beginnings of our own Ordnance Survey as well as its sibling Military Survey.

Postscript – As this issue is about to go, I have to record that the area of bushes mentioned above has been cleared and is being included with a much larger area to help satisfy the insatiable demand of Heathrow for car parks. However, I was assured that the contractor's instructions are to 'work round the cannon' (CCS members keep your fingers crossed).

Sheetlines 50, December 1997

Brian's manifest interest in the baselines of the Ordnance Survey was further demonstrated in the pages of Sheetlines 62, where the photograph opposite appeared of him at the southern end of the Hounslow base. Peter Haigh, in his article 'The bases of the Ordnance Survey' (pages 45-56 in the same issue), paid due acknowledgement to the co-ordinate data supplied by Brian, adding 'Members may also be interested to note that Brian Adams's rabbit, or at least one of its descendants, is still to be found'.

Brian Adams (left) and David Watt, in 2001, inspecting the Hampton Poor House Cannon which marks one end of Roy's Hounslow Heath base. Brian's transcription of the plaque, below, appears on page 26.

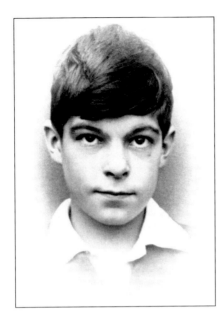

Brian as a child, left, and many years later with his wife, Mabel, at the 1967 christening of their first grandchild, Mark.

At the controls of a Kongsberg flatbed chart plotter, date unknown.

Brian was a benevolent, and very welcome, background presence at many Charles Close Society meetings. Sometimes he could be persuaded into the limelight as (above) giving the address at the 1994 AGM and (below) accepting honorary membership from Chris Board in 2005.

From eighteen minutes west to longitude zero
Episodes from the lives of a cartographer and a meridian

Brian Adams's talk given before the Annual General Meeting at Birkbeck College, on Saturday, 14 May 1994, with some adaptation to appear in print; however much of it remains addressed directly to the reader.

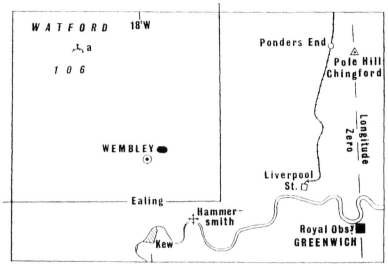

Figure 1

It is an ancient Mariner,
And he stoppeth one of three.

S.T. Coleridge

I shall come to Coleridge later; meanwhile at *figure 1* is a simplified map, based on the Ordnance Survey Popular Edition with some special symbols. The dot and circle indicates the blue plaque (not yet installed) 'Brian was born here', the event occurring sixteen months after that of the other local celebrity also in action today.[1] It was here, in longitude 0° 18' 00" West, as I hoped you guessed, that I started looking at maps at quite an early age, and fairly soon drawing maps. I cannot really put a date to this beginning, but very early on I had tucked in to the pages of my *Comparative Atlas* freehand pencil maps of the home countries and the continents, one feature of which was that the small countries were drawn bigger so you could see them better; this already typified my life long approach in that I did not leave them off, I made sure they were on. My ploy was part of the Art of cartography, the Greenwich Meridian shown across the map is part of the Science of cartography, and I make a serious point here that both these facets are important and neither should ever be forgotten;

---

[1] Wembley Stadium; the 1994 AGM was held on FA Cup Final day.

this is something the electronic cartographers in particular should keep in mind, even in such a basic operation as the placing of names. Having mentioned electronics, I interpolate here that my piece excludes the last ten years or so, and some of my remarks may not strictly apply today.

My geographical education also began early; both my parents were Fenfolks, so we used the line out of Liverpool Street regularly, and one of my first questions was "Where is Ponders Beginning?", to which I never received a satisfactory answer! One of the first things I remember being taught, though I do not remember why, was '*J*illingham in Kent, but *G*illingham in Dorset', but it was not until I was working with a Geordie that I was instructed of the difference between Belli*ng*ham in metropolitan Kent and Belli*nj*am in Northumberland; this is a reminder that you cannot fully comprehend the country simply by looking at a map of it, though anyone with a good eye for country can get a good visual impression of it from a decent map. Here I submit that the older maps, the New Series or Popular Edition say, give a better view than the modern ones do; there is no doubt that the human cartographer put something indefinable, instinctive, into his or her drawings that the machine simply cannot.

It was not long before I was bought my first Ordnance Survey map, 106 *Watford*, purchased on a shopping trip to Ealing, one of the 90% of all known places lying across the join of two sheets (that is a figure mathematics cannot explain), and showing single line connections from Bushey & Oxhey to Croxley Green Junction and Watford High Street;[2] this I believed implicitly at the time although my father said it was wrong, and I only fully accepted it to be an error when we passed a train going the other way there. So I learnt fairly early on that the Ordnance Survey could make mistakes, then we purchased 85 *Cambridge*, only to find that my mother's village six miles north of Cambridge was off the map; we therefore got the adjoining sheet, 75 *Ely*, which also included my father's village, but this had no colour on the stations and showed the Great Eastern Railway! We next bought 114 *Windsor* showing the other half of Ealing, as well as the favourite excursion spot of Kew Gardens, and thus I was learning fast about Ordnance maps whilst continuing to draw maps of all sorts myself, and when, at the end of term, we were told to take a book to read while the teacher was marking papers, I would take a couple of maps to look at.

By then I felt well qualified to undertake my first major project, the Ordnance Survey of 'Doggy Land', mapped as you might guess at one inch to a mile with some town plans, the first of which was drawn on what is known technically as a non-permanent medium; the reader might call it the steam on the kitchen window. I realise now that at that tender age I was unaware of the principles of triangulation, so the maps can not have been as good as I thought they were; I was also unaware of the provisions of the Copyright Acts, so, as like all the best Ordnance Surveys all the records were lost during the war, it is no good going round to the British Library and expecting our batch of members there to turn out copies. I am hoping this confession does not render me liable to prosecution, but having recently constructed some index maps for the British Library I am trusting that honour is more or less satisfied!

I now move forward to a time when three of the more senior members of the Charles Close Society were all at school together – our long-serving first chairman, Peter Clark, first secretary and past-president of the British Cartographic Society, Ian Mumford, and myself.

---

[2] Shown at **a** on *figure 1*; these curves were always double track.

Ian Mumford and I started on the same day, Peter Clark a year later. I joined Ian in 2A after one term, but I had started in 2B under a gentleman called 'Slogger' Logan, and on that first day Slogger informed 2B "I remember when I taught the Astronomer Royal in this very room", and I have been somewhat hooked on the Greenwich meridian ever since. The Astronomer Royal in question was Sir Harold Spencer Jones, who subsequently took the observatory to Herstmonceux and put it on atomic time. The school was Latymer, Hammersmith, marked on *figure 1* by a cross crosslet, sometimes known as the Latymer Cross, there being fourteen of these on the full Edward Latymer coat-of-arms.[3] While I was at Latymer we had a new master, one 'Doc' Briault, and whilst he taught us Geography, from time to time I taught him about Ordnance Survey maps!

Readers familiar with my writings should realise that the sides of the map at *figure 1* are parallel to the meridian of Delamere, 2° 41' 03".562 West from Greenwich. The much missed Guy Messenger[4] took me to task, saying that such an accurate position was meaningless on a one-inch map; with that I agreed, but I explained to Guy that I was not quoting a position on a map but that of the triangulation station which was specifically the origin of the projection and co-ordinates used to construct most medium and small scale Ordnance Survey maps before the adoption of the National Projection. However, for my present purpose it will be sufficient to say that Delamere is 2° 41' west of Greenwich, whence Greenwich is 2° 41' east of Delamere; but on the map the Greenwich meridian runs at an angle of 2° 06' to the side.[5] This is known as the convergence of the meridian,[6] and it is specific to the map and the place being considered; were the map on a different projection the angle would be different, whilst if we follow the Greenwich meridian on the Popular Edition from the area of *figure 1* up to the Shetlands the convergence increases to 2° 20'.

I feel I now need to insert a brief word on academic Britain in wartime, not so much forgotten as not remotely thought about; the evacuation of school children is fairly well known, but most of London University was also evacuated, one college being scattered all round the Welsh coast, whilst three colleges and a medical school had the great good sense to go to Cambridge. Here, as in every university town, academia, especially male academia, was very tightly controlled by a body called the Joint Recruiting Board; you had to get their permission to go to college in the first place, they controlled what you did when you got there, and where you went when you finished. So it was that I became a London University student in Cambridge, thereby getting the best of both worlds, and came under the aegis of the Cambridge Joint Recruiting Board and particularly its chairman 'That man Snow'; you may know and maybe like his work as C P Snow, but you will gather that we did not like him at all. Now because of the way warfare was developing, Snow issued an edict that all male students in certain disciplines had to do a Radio course; the stupid thing about this was that those of us in our final year just did the first year of the Radio course. Then, having sat finals but not knowing the results, we all passed before Snow and a couple of his acolytes and were classified as for Research, Forces, or Industry (which included everything else), and those including myself designated for 'Industry' had our names passed to the Ministry of Labour to find us jobs; this was the way things were done in wartime.

---

[3] North London members will know of another Latymer School at Edmonton; it was a separate foundation under the same will of Edward Latymer, who died in 1624.
[4] Every main speaker at the AGM referred to the loss of this revered Honorary Member.
[5] Any divergence from this angle on figure 1 is due to the photocopier; the slope is correct on the original drawing.
[6] Not to be confused with convergency, which is the angle in space between the meridians at two points on the earth's surface, an angle used in some navigational calculations.

Having been sent for interview to a number of radio firms and turned down, because they all took the same attitude "We could give you a job on the bench making radio sets, but it wouldn't be fair to you", the Ministry of Labour wrote to me to say "Perhaps you would like a mathematical job"; what I said and what I wrote back were in two entirely different languages but they both meant "Yes", and so I was sent for an interview at the Ministry of Aircraft Production for a job in the middle of Salisbury Plain, which was a major centre of the war effort. However, on the morning I was going for this interview a letter came, forwarded by my college, saying that the Hydrographic Department wanted a mathematician for an Admiralty Cartographer. My mother looking over my shoulder said "Sounds perfect, doesn't it?"; indeed it did sound perfect – I had been drawing maps all my life, and these people wanted to pay me for doing it! £266 15s a year, which was quite a goodly sum at that time. So, having been offered the post on Salisbury Plain, I wrote to the Ministry of Aircraft Production to say that I was interested in this other job, adding very naively but for once apparently successfully that, because cartography was something I was very interested in, I should be even more use to the country doing that than their job, sat back, and nothing happened. Eventually I wrote to the Hydrographic Department, who wrote back and said "We asked the Ministry of Labour if you were available, and they said 'No'". So I told them what I had done and that I had heard nothing to the contrary. They interviewed me and accepted me on the spot, and after I had been working there for seven months, I received a letter from the Ministry of Labour, addressed to me at the Hydrographic Department, giving me permission to take the job! I am not alone in sometimes wondering which side the Ministry of Labour was on?

I should explain that at that time the Hydrographic Department employed just one professional mathematician, and it was because he was due to retire that they had advertised for another one; although I was taken on to replace him, the only time this person ever spoke to me was when he came round to say 'good-bye'. However, at the same time the Department was employing a totally amateur mathematician who was far more active, far more use to the Department than the aforementioned professional. This was J C B Redfearn, and I tell you that the initials 'JCB' represented a power in the Hydrographic Department long before Mr Bamford started making his machinery. Also, this was the period between D-day and VE-day, and you will understand when I mention it that, as opposed to the military cartographers, the Hydrographic Department's stint of 24 hours a day, seven days a week activity was largely over once the Normandy landings were completed. Hence they were already turning their attention to the Far East, and starting a programme of gridding or re-gridding the charts of strategic ports and coasts, Chinese rivers, and so on. And so, although as the mathematician I was destined for the section dealing with special charts, navigational diagrams and geodetic records, on my arrival at the wartime Admiralty in Bath, I was grabbed by the Far East section to assist in their gridding programme. The important thing about this was that the then Deputy Chief of the Far East section was the aforementioned J C B Redfearn, and so, not only was I able to learn the mathematics of surveying and cartography from *the* expert, he had me literally at his elbow where I could advise him and check his more advanced developments, which he had previously to send to others outside the Department.

Now the Charles Close Society is for the study of Ordnance Survey maps, Ordnance Survey cartography; I was a hydrographic cartographer, so is there a difference? On the mathematical side not much – surveying computations are exactly the same, and although we

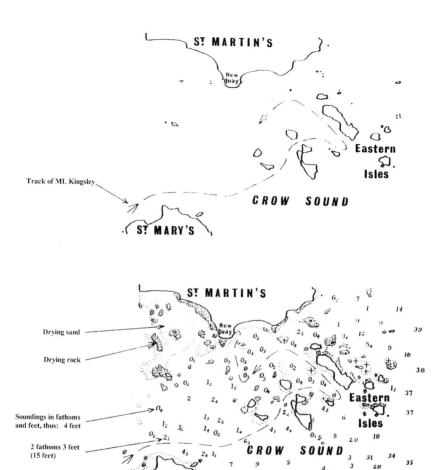

Figure 2: map (a) above and chart (b) below

use different projections, which means different formulae or different tables, the calculating techniques are essentially the same (and I fear the results would be all Greek to many readers); drawing techniques are the same, but what of the content? You may have read that a map of the sea is called a chart; however, there is rather more to it than that. Let me take you back to a holiday in the Isles of Scilly, and the morning I took the launch to St Martin's. *Figure 2a* shows the route we followed as we cruised into Crow Sound, through the Eastern Isles to see the baby seals and sea birds, and then turned sou'west; I was on my favourite perch on the stern and a lady sitting next to me said "I thought we were going to St Martin's"; I said "We are.", and she replied "Why are we going away from it then?" I have to admit a fair question for a landlubber, we were going away from St Martin's; but *figure 2a* is the map and it depicts what you can *see*: *figure 2b* is the chart; it depicts what you *cannot* see, and without going into the niceties of hydrography I said to the lady "There's only a foot of water across there." "Oh, I see" she said, adding "It's a good job he knows that." Well, I am sure that the reader is aware that underwater dangers abound all around the Isles of Scilly, and no boatman would last an hour if he did not know what was below the surface.

Some explanation of the chart symbols is given on *figure 2b*, which belongs to the pre-metric era; the soundings show the depths of water below the level of Low Water Spring Tides, so that at any given time the actual depth of water is the sounding *plus* the height of the tide above that level. The firm coastline is the line of Mean High Water Springs, and the hydrographer uses the expression 'drying' for any feature between High and Low Water Springs, although at Neap Tides it may never dry out or may never get wet, if it lies near the bottom or top of that range respectively. The cross symbol for an underwater rock (not an underwater church, despite appearances!) is the oldest hydrographic symbol, dating at least back to the fifteenth century. The reader will observe that, even without venturing across the one foot patch, the voyage of *Kingsley* that morning involved some very accurate manoeuvring, local knowledge being essential.

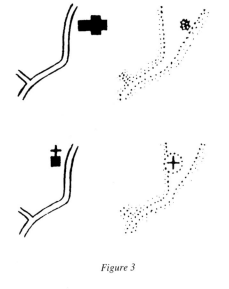

Figure 3

Some time before that holiday in Scilly, one of my colleagues wrote to the Ordnance Survey to tell them that they showed a church on the wrong side of the road, and they wrote back and thanked him. But if you were going to church, even an unfamiliar one, on a Sunday morning, you would not go into the field on the left (see *figure 3*) because the map marked a church there; you would go into the ecclesiastical-looking building on the right. However, if the patch of underwater rock were charted on the wrong side of the channel, a ship would steer to the eastern side, have her bottom holed, and the crew and passengers would be lucky to escape with their lives. Yet such an error can arise from so simple a fault as putting a tracing down the wrong way up, and this is the reason for the checks and procedures which have arisen during two hundred years of Admiralty charting. The reader should also be able to see the obvious need for such a danger to be charted as soon as its existence is discovered, which is why chart plates are corrected continuously, charts already printed are corrected daily by hand from Admiralty Notices to Mariners, and ships at sea are notified of serious new dangers by radio navigational warnings, supplemented by weekly editions of Notices to Mariners. And I add that this awesome responsibility of not endangering lives does not lie upon Ordnance Survey cartographers, let alone Mr Bartholomew's and others'.

To return to our Sunday morning walk, a single glance at the map would tell you exactly where you were when you passed the side turning or the bend in the road; but you cannot tell from looking over the side of the ship that you have just passed the swatchway or that it is now that you must swing to port. So to find out where you are on the chart you have to fix your position by angles or bearings to objects ashore, or maybe by a Decca Navigator fix.[7] You then know that you are, say, just at the turn in the channel and steering 015°; but the tidal stream is running at 240° and there is a gale blowing on your starboard beam, so which direction are you in fact going? You cannot and must not guess; that dangerous rock is ahead of you and you have to take another fix; what this amounts to is that the chart is part of your navigational set, a navigational instrument in its own right, and this is the vital difference between a chart, marine or aeronautical, and a map.

$$E = \lambda v \cos \phi + \lambda^3 \frac{v}{6} \cos^3 \phi \left(\frac{v}{\rho} - \tan^2 \phi \right)$$

$$+ \lambda^5 \frac{v}{120} \cos^5 \phi \left\{ 4 \frac{v^3}{\rho^3} \left(1 - 6 \tan^2 \phi \right) + \frac{v^2}{\rho^2} \left(1 + 8 \tan^2 \phi \right) + 2 \frac{v}{\rho} \tan^2 \phi + \tan^4 \phi \right\}$$

$$+ \lambda^7 \frac{v}{5040} \cos^7 \phi \left(61 - 479 \tan^2 \phi + 179 \tan^4 \phi - \tan^6 \phi \right) + \ldots$$

$$N = m + \lambda^2 \frac{v}{2} \sin \phi \cos \phi + \lambda^4 \frac{v}{24} \sin \phi \cos^3 \phi \left(4 \frac{v^2}{\rho^2} + \frac{v}{\rho} - \tan^2 \phi \right)$$

$$+ \lambda^6 \frac{v}{720} \sin \phi \cos^5 \phi \left\{ 8 \frac{v^4}{\rho^4} \left(11 - 24 \tan^2 \phi \right) - 28 \frac{v^3}{\rho^3} \left(1 - 6 \tan^2 \phi \right) \right.$$

$$\left. + \frac{v^2}{\rho^2} \left(1 - 32 \tan^2 \phi \right) - 2 \frac{v}{\rho} \tan^2 \phi + \tan^4 \phi \right\}$$

$$+ \lambda^8 \frac{v}{40320} \sin \phi \cos^7 \phi \left(1385 - 3111 \tan^2 \phi + 543 \tan^4 \phi - \tan^6 \phi \right) + \ldots$$

Figure 4: 'Redfearn's formidable expressions'

Going back to things in the Hydrographic Office itself, it was Redfearn's preoccupation in the late forties to persuade the department to adopt the Transverse Mercator Projection, not then for its charts but for its surveys, because it was his contention, quite rightly, that the very basic methods used by the Naval surveyors for their computations at that time totally hid the quality of their observation work. It was in the course of compiling a handbook to expound the benefits of the Transverse Mercator Projection to the Naval Surveying Service that he became dissatisfied with some of the formulae which the military men had developed for that projection, and he delved deeply into them himself. He duly developed what Derek Maling has described as 'Redfearn's formidable expressions ... from his classic paper on the Transverse Mercator', which have a particular significance to the Charles Close Society as the formulae for the National Grid. Seen in the illustration (*figure 4*) [8] are those for the conversion of latitude(φ) and longitude(λ) to easting and northing, and there is a fairly similar pair for the conversion of rectangulars (easting and northing) to geographicals (latitude and longitude), with others for the calculation of convergence, scale factor, etc. They are what is known as 'convergent infinite series', which means that in theory they go on for ever with

---

[7] There is no buoyage present which might have assisted positioning (it may have been removed for the winter ice-up). Also, I remind readers that I am writing of the days before GPS.

[8] I did say they were all Greek!

increasing powers of λ, but they are cut off where shown because the terms get rapidly smaller in value despite seeming bigger on the paper; the last terms shown in the formulae are already near enough zero.[9]

My involvement in these formulae was, firstly to check Redfearn's workings thus far (and there was a minor arithmetical error in one of the complex terms), and then to derive two more terms in each expression in order to make absolutely sure that nothing was being ignored which could possibly affect results than by more than a few millimetres (on the ground that is, not on a map). Redfearn sent my workings to Harry Brazier of the old Colonial Surveys for checking, checking being a vital ingredient of any mathematical process (and, yes, there was a minor arithmetical error in one of the complex terms). What I really want to bring home to you is that these formulae, these 'formidable expressions', were derived by a man who had no formal mathematical education beyond elementary school, which in his day meant age fourteen, rather below today's GCSEs. Redfearn taught himself these higher reaches of mathematics simply because, just like all CCS members, he was interested in maps and all that went into them. So reader, you could do it too; you doubt it? – well, you might surprise yourself.

Members are certainly aware that to map the curved Earth requires a projection, but I cannot emphasise too strongly that the actual process embodies the use of the projection formulae; without the projection, without these formulae, there could be no triangulation, no survey, no sheet lines, and with no survey and no sheet lines your Ordnance Survey map would be a blank sheet of paper, and the Society would not have lasted thirteen years discussing blank sheets of paper! So when you see the mention of a projection in *Sheetlines* please don't knock it, it is vital to the actual existence of the map you are interested in; turn on if you must, but maybe glance through the item twice, even thrice, and hopefully get the general idea, and if you would like to discuss it my address is in the membership list.

Before leaving the projection formulae I tell you that the Hydrographic Department, in other words Redfearn, used them to produce their own National Grid conversion tables a year before the Ordnance Survey published theirs. Meanwhile, it was about this time that the OS began to issue the National Grid co-ordinates of retriangulation stations on a regular basis, and Redfearn's natural inquisitiveness led him to compute some of their geographical positions to compare them with their positions in the old Principal Triangulation; he was probably the only person in the country outside the Ordnance Survey capable of doing so at that time. He found that there was a fairly regular difference between the old and new positions of around 2¼ metres, but an enquiry about this directed initially to the Astronomer Royal[10] elicited a much more startling fact, which I now explain from square one. The Royal Observatory at Greenwich was founded in 1675 for the specific purpose of 'finding the so-much-desired longitude of places for the perfecting the art of navigation', in other words, so that British ships carrying British trade around the world could find out exactly where they were. The actual problem of 'discovering the longitude at sea' was finally solved by John Harrison in 1759, but an essential component in the matter was the installation of Bradley's transit instrument in 1750, superseding a less permanent transit set up by Halley in 1721.[11]

---

[9] These final terms have already been simplified by substituting the close approximation of unity for the ratio v/ρ; ρ, v are the Earth's radii along and perpendicular to the local meridian respectively.

[10] The Astronomer Royal was then ultimately responsible to the Hydrographer of the Navy, though not in the Hydrographic Department proper.

[11] Halley (1740-42), Bradley (1742-62), Pond (1811-35) and Airy (1835-81) were Astronomers Royal; Spencer Jones was in office 1933-55.

I may need to explain exactly what a transit instrument is – it is a telescope so mounted that it can only rotate in the north-south meridian, and it is used to time the passage of heavenly bodies across that meridian; from such timings differences of longitude can be determined, and this is why the geographical positions of observatories listed in the Nautical Almanac or elsewhere are the positions of their transit instruments. In the particular case of Greenwich, the meridian through the observatory transit is the meridian of zero longitude. One result of the particular construction and use of the transit telescope is that it does not have the familiar dome above it, but a simple slit in the roof and upper walls of the building housing it. So Bradley's transit marked longitude zero until 1816, when it was replaced by Pond by a bigger instrument on the same mounting; then in 1850 G B Airy, later Sir George, decided on some reconstruction at the Observatory and in the following year erected a further new instrument in a different situation (see *figure 5*). In those days many countries had their own zeros both for longitude and for time, and to eliminate the confusion caused by this practice an international conference was held at Washington in 1884, which decided that henceforth Greenwich, that is the Airy transit, would be the international zero.

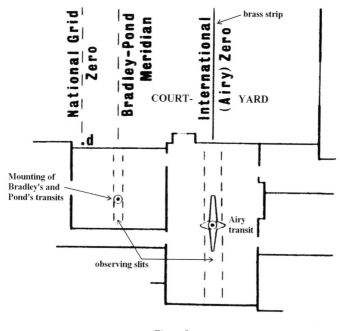

Figure 5

The attentive reader will have remembered that three years ago we were celebrating the bi-centenary of the Ordnance Survey, which grew from an earlier operation to connect the observatories of Paris and Greenwich, which takes us back a long, long time before 1851. So indeed, Roy's original triangulation was connected to Bradley's transit at Greenwich, and the later observations of the Principal Triangulation were made into Pond's on the same site, and it was from the accepted astronomical position of that site that positions were derived throughout the old triangulation. The final calculations of that triangulation were commenced by William Yolland in the 1840s, but were completed by A R Clarke in the late 1850s and published in 1858, seven years after the installation of the Airy transit. Whilst there were good reasons why Yolland's inevitable use of the Bradley-Pond meridian as the zero should have been retained in Clarke's 'Account', it is very much a mystery why Clarke made no reference whatsoever to the new instrument, not even in a postscript. His silence on that point probably contributed to the fact that when the Airy transit was first connected to the Retriangulation in 1949, it came as a total shock to the Ordnance Survey that its longitude was found to be not zero but 0.4 second East, equivalent to a displacement of about eight metres.

An enquiry into this revelation elicited the facts which I have related above and a report by Colonel Shewell,[12] one of the Survey's more responsible officers, concluded that 'the discrepancy cannot and should not be covered up', but his superiors did nothing to tell anyone about it until Redfearn discovered it for himself two years later. But the forgotten change from Bradley-Pond to Airy only accounted for about 5¾ metres, leaving an apparent error in the Retriangulation position of the Airy transit of 2¼ metres, the same discrepancy that Redfearn found between old and new positions in the area. To understand this I need to explain something about the adjustment of a triangulation, and I hope the reader will continue to bear with me in this. Why does a triangulation need adjustment? – mainly because it involves observing objects 30, 50, even 100 miles away through the atmosphere, and the atmosphere does very funny things to rays of light, especially near the ground, so that a ray from a distant point does not reach your theodolite at the angle it started from.

My diagram *figure 6* shows what is termed 'Figure 1' of the Principal Triangulation, and as I have explained all the observed angles are wrong, or shall we say 'not quite right', and

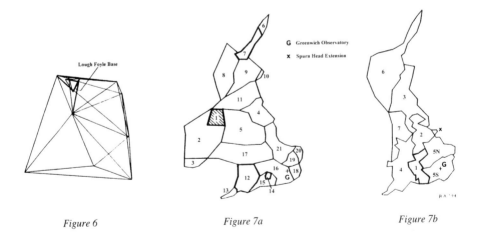

| Figure 6 | Figure 7a | Figure 7b |

---

[12] Subsequently Brigadier.

they are not consistent with each other. What we have to do is to determine corrections to all the angles to make them all consistent, with as little disturbance to the original observations as possible; the result may not be absolutely correct, but it will be as near as it is possible to get. The computation involves simultaneous equations – perhaps you remember them from school, a set of three, four, or more? But this is not school maths, it is not university maths, it is real life maths; to adjust the Figure 1 we have to solve 39 simultaneous equations, and this means working through 265 equations altogether. Having gone through those 265 equations we have only dealt with the area shown hatched in my *figure 7a*, which depicts all the adjustment 'Figures' of the Principal Triangulation, and you will realise at once why the whole network could not be adjusted in one go. But turning now to the Figure 2 we have to leave undisturbed the sides we have already fixed in the Figure 1 adjustment; this imposes a small constraint on the adjustment of Figure 2 which is slightly less satisfactory than that of Figure 1, and the more figures we adjust, the more constraints we have to impose on those which remain. The figures outlined with heavy lines were adjusted totally freely, without any fixed sides, and Figures 1 and 14 included the Survey's two most accurate baselines, the only ones included in the final adjustment.

Having adjusted the whole network, only then can we begin to find out where everywhere is. As I said earlier, the astronomers have given us their position of the Pond Transit, and the astronomical bearing of the ray thence to the Chingford station on Pole Hill;[13] using these data plus the distance Greenwich-Chingford from the adjusted triangulation we obtain the position of Chingford. Then from the triangle Greenwich-Chingford-Leith Hill we obtain the position of Leith Hill, from the next triangle we obtain that of Butser at the summit of the South Downs, from Butser to Dunnose on the Isle of Wight, thence to the Ordnance Survey Office at Southampton, so that they knew where headquarters was; also from Dunnose, Black Down in Dorset was fixed and three more hops arrived at Goonhilly on the Lizard;[14] and in a similar fashion positions were fixed all over the British Isles. But because the adjustment of the Principal Triangulation had started from the west, the biggest constraints on the individual figures occurred in the east, and there was a slight distortion of the position of Greenwich within the finalised geometric network. Consequently, working back from Greenwich, most of the geographical positions obtained in the central and western parts of the British Isles were slightly in error.

Moving rapidly forward to the current century, the overall adjustment of the Retriangulation, which was confined to Great Britain, was done similarly to that of the old triangulation, although its figures were larger and therefore fewer (see my *figure 7b*).[15] However, a quite separate adjustment was also made using eleven stations which were occupied in both the old and new triangulations; by this, without disturbing the adjusted shape of the new network at all, it was laid down in the position which gave the best overall

---

[13] On this hill due north of Greenwich (see my *figure 1*) Pond erected a meridian mark which remains there today, together with a retriangulation pillar; the latter is within a few inches of the Airy meridian. The Principal Triangulation station was a buried mark, lying 13½ feet almost due south of the meridian mark.

[14] Well before modern technology made Goonhilly a household name I walked across these Downs, feasting off the thousands of blackberries with a total absence of other human beings. It is salutary to recall that over a hundred years earlier, but in mid-winter not blackberry time, the area had been occupied by an intrepid party of Ordnance Trigonometrical Surveyors.

[15] Work on Figure 5 was halted by the outbreak of the Second World War, but the need for good positions around London led to its southern half being adjusted in its entirety, with the northern part following.

fit with the old work.[16] Therefore the geographical positions of the new work were infused with the slight error obtaining in the central area of the old, and again, working out to Greenwich through the adjusted Retriangulation, a slight additional distortion was brought in. The sum result of all this was the 2¼ metre discrepancy we have been searching for, and for the physical meaning of this I refer you back to my *figure 5*. The National Grid longitude of the Airy transit, the international zero meridian, is 0".42 East, and that of the Bradley-Pond meridian is 0".12 East; this means that the line of zero longitude in the National Grid runs 0".12, or 2.3 metres, west of the old Pond instrument, and if you measure it out on the ground you will find that the National Grid zero is marked by a drainpipe **d** on the plan.

One more point, you may have read or been told that meridians and parallels are imaginary lines on the Earth's surface, but some of you know better, I am sure; you have seen the Greenwich Meridian running across the observatory courtyard, and have probably stood with one foot East and one foot West, and if you did, and looked up, you would have seen the observing slit in the wall I mentioned earlier, with the small opening roof. The slit over the Pond instrument has been glassed in. Or you may have seen the meridian line running across the sea-bank near Cleethorpes or on the cliff top at Peacehaven; but if you have seen it at Cleethorpes or Peacehaven it was the Bradley-Pond meridian and not the Airy, and most of the other meridian markers of various types are on the Bradley-Pond meridian, because they have been installed using Ordnance Survey data. But in 1984, the centenary year of the Washington conference, the Ordnance Survey offered to mark your house if the zero meridian ran through it, not to mark the actual meridian, just the house, and I am assured that for this operation they made the necessary correction to ensure that they were dealing with the international Airy meridian.

Finally, reader, I refer back to the quotation at the head of this article to remind you that it was the needs of the Ancient Mariners which led to the founding of the Greenwich Observatory in the first instance, but if you refer to the Greenwich Meridian today, it is One of Three!

Sheetlines 40, August 1994

---

[16] In the course of proposing a vote of thanks for the talk at the AGM, Ian O'Brien remarked that this somewhat controversial procedure was a typical British compromise, introduced by the architect of the Retriangulation, Major (ultimately Brigadier) Martin Hotine. Hotine, subsequently Ian's chief, was also self-taught in the higher realms of mathematics.

In defence of Mudge (and also myself)

My contribution to the visit to the Public Record Office last March was relatively short, and the reference to it in the report in *Sheetlines 61* was equally so. As a result it was somewhat misleading and, especially as I received a letter from a concerned non-member of the Society only a week after *Sheetlines* dropped into my letterbox, I hope I may be able to explain matters further to those who were not present at the PRO. Briefly, I showed that the latitudes towards **eastern** Kent were wrong, and they were not really Mudge's anyway. I remark on two distinct matters.

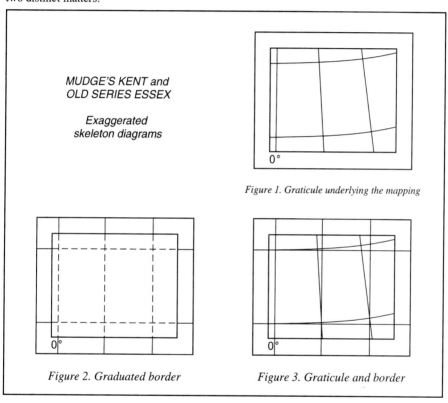

MUDGE'S KENT and
OLD SERIES ESSEX

Exaggerated
skeleton diagrams

Figure 1. Graticule underlying the mapping

Figure 2. Graduated border

Figure 3. Graticule and border

The graduations on Mudge's Kent and Old Series Essex

Kent lies between Sussex and Essex not only geographically but also historically cartographically. Gream's (or Gardner, Yeakell and Gream's) Sussex, published by Faden 1795, was a fairly traditional private venture county map, but it was completed with some assistance from the new ordnance surveyors; it crossed the county boundary in just a few places. Mudge's Kent was surveyed and drawn under the Board of Ordnance, but was published by Faden in 1801; it included a substantial area of Essex right up to the north border, Middlesex to the north and west borders, moderate areas of Surrey and Sussex, and hydrogaphy in the Thames Estuary east from Southend. Then came *Part the First of the General Survey ... Essex*, published by Mudge at the Tower of London, 1805, the first 100% Ordnance map, which was completed out to all the landward borders and also included the outer Thames Estuary hydrography.

Each of these counties was engraved in four parts, Sussex in four portrait sheets, the others in conventional quarters, and each when mounted complete was surrounded by a graduated border. That around Sussex was distorted, as may be found round a number of earlier county maps and the full reasons for which require further investigation, but which presumably attempt to reconcile the graduations with positions obtained by various methods. But Kent and Essex both have graduations of an entirely different type, which if completed across the maps form rectangular patterns with equally spaced meridians and equally spaced parallels. These are technically on the Modified Plate Carrée Projection, and are illustrated in skeleton form by figure 2.

However, these two counties had been mapped on the basis of the new national triangulation, and were drawn on something close to Cassini's Projection, which is demonstrated in exaggerated form in figure 1. (They were not strictly on Cassini's as the triangulation had been adjusted without introduction of spherical excess.) The Greenwich zero meridian ran close to the western edges of both counties, so it was fairly natural that it was chosen to be the central meridian of the projections. Similarly, it was the initial meridian for the graduated borders.

Figure 3 demonstrates the effect of the two projections superimposed, showing that geographical positions (latitudes and longitudes) taken from the graduated borders will be correct along the Greenwich meridian but will be increasingly in error the farther they are to the east. Whilst the effect on longitudes varies due to the sloping meridians on the true mapping projection, the error in latitudes increases progressively eastward, culminating, for example, in the true parallel of 51° 20' meeting the eastern border of the Kent map some 8 mm or 16.5 seconds north of the corresponding border graduation tick. Direct comparisons may be made between the projections using the 5' intersections printed on modern one-inch maps and the equivalent intersections plotted from the original graduated borders and transferred by local fit. Remarkable consistencies among the offsets between pairs of points brought out by this method serve to demonstrate the high quality of the surveys carried out by Mudge and his fellow ordnance surveyors.

The reason for the combined use of this extraordinary dichotomy of projections can only be guessed at after the considerable lapse of time, but no guess seems to make any sense at all. The nascent ordnance survey comprised quite a small body of professionals working together, whose leaders were certainly mathematically competent, and who would have been in no doubt as to the true positions of detail. If the Kent border had been added by an over zealous engraver, the error would surely not have been repeated on the Essex map. Could the two borders have been imposed by some outside authority? "Pass"! In any case there seems to be a rebellious undertone in the note on Old Series sheet 1, reproduced as figure 4. I have

The Scale of Latitude and also that of Longitude around this Map, being drawn & graduated on a plane projection,

the Latitudes & Longitudes deduced therefrom, can be only nearly true near to the Meridian of Greenwich.

Figure 4. Note engraved (in one line) below the border of Old Series sheet 1, Essex south-west.

been asked more than once what this means; taken at its face value it seems fairly meaningless, but it must have been intended as a warning against taking positions from the graduated borders for the reasons explained above. Hence, given this significance, it should also have been applied beneath Mudge's Kent. It only remains to remind readers that after the publication of Old Series Essex graduated borders disappeared from one-inch maps for many years.

Various states of Mudge's Kent

The visit report in *Sheetlines 61* also remarks that my copy of Mudge's Kent was an early state, repeating a comment by one of the party based on the fact that my map did not show the Royal Military Canal. Further investigation suggests that, following Dr Joad (of the radio *Brains Trust*), "it depends what you mean". Volume I of Margary's Old Series reproductions lists sixteen copies of Mudge in which can be found the following number of states, sheet by sheet: NW sheet 4, NE sheet 2, SW sheet 4, SE sheet 3. These can be found in seven different combinations in the complete maps, which are, listing the sheet states in the same order, 1111, 1121, 1122, 2131, 3232, 4242 and 4243.

Most of the corrections to the plates between states were fairly minor, the only major correction being the insertion of the Royal Military Canal on the SE sheet, which only appears on the final third state. Hence it only appears on the seventh and last of the combined states, and from one point of view all the first six combinations were early states. However, my own copy is the same as the fifth of the seven listed above, so it is not among the earliest. Meanwhile, Peter Clark and myself, and probably others, cannot but wonder whether any further copies of Mudge's pioneer map have emerged during the past twenty-six years.

"Parallel to the Meridian of Butterton Hill" – do I laugh or cry?

There was something rather comforting about the inscription "Parallel to the Meridian of Greenwich" found against the eastern borders of a number of the Old Series one-inch maps in southern England, from Kent to Hampshire, and the meridian line itself engraved across or outside the north and south borders of a column of these maps extending from Sussex to the East Riding. This suggested that Britain's first comprehensive series of reliable maps was firmly based upon Britain's own prime meridian, that of its Royal Observatory, which was subsequently adopted as the international zero meridian at a Washington conference in 1884. But the big drawback about Greenwich was that it was nowhere near the north-south centre line of Britain and would therefore have been most inconvenient as central meridian of a national map series, and so the sheet lines of Old Series Part II, Devon, were made parallel/perpendicular to an unspecified west country meridian.

For reasons which I detail in a technical section below, the early Ordnance topographical survey was computed in several portions centred on different triangulation stations across southern England, and in the area of Devon it was referred to Butterton Hill (near Ivybridge) and its meridian of longitude. The line of this meridian, from which base all the map detail was drawn in, was engraved in the southern margin of sheet 24 and the northern margin of sheet 27. Parts III and IV of the Old Series abutted Part II on its west and east sides respectively, and all the sheets in these parts had east or west borders labelled "Parallel to the Meridian of Butterton Hill", on the evident assumption that the borders of Part II had been made parallel to that meridian.

But what no-one, myself included up to now, seems to have noticed is that the meridian of Butterton Hill, as indicated by the markers on sheets 24 and 27, runs across Part II at an angle to the east and west sheet lines. Consequently, neither these lines nor any of those in Parts III and IV were parallel to that particular meridian, whilst those in Part IV were not even parallel to those in Parts II and III (apart that is from the western borders of sheets 17 and 18 which were common with the eastern borders of sheets 21 and 22).

As I conclude in the technical section, the sheet lines of Parts II and III were probably made parallel/perpendicular to the meridian of 3°W, whilst those of Parts I, V, VI and VIII, excluding sheet 10, were similarly related to 0° (Greenwich). This left Parts IV and VII to be fitted in between converging sets of sheet lines, with the results that all their sheets were both narrow and tapering. This is the reason for the lack of parallelism in Part IV referred to above, and was also responsible for the two columns of narrow, tapering sheets stretching right up to numbers 88 and 89 (sheets 11 and 13 were also narrow with 45 to 87 narrow and tapering).

I am at a loss to understand how the labelling error could have come about in such a compact organisation as the nascent Ordnance Survey, especially as the map detail was correctly laid down in relation to the true orientation of the sheet lines. But, whilst my immediate reaction to Parallel to the Meridian of Butterton Hill is now: "Don't make me laugh", it must be considered tragic that this mis-description has gone unnoticed for so long, and has been taken as gospel by those relatively few writers who have attempted to investigate the construction of the Old Series maps. One further thought: what if there had been Trading Standards Officers in 1813?

THE OLD SERIES ONE-INCH
south of Preston – Hull
(omitting quarter-sheet lines)
showing PARTS I to X

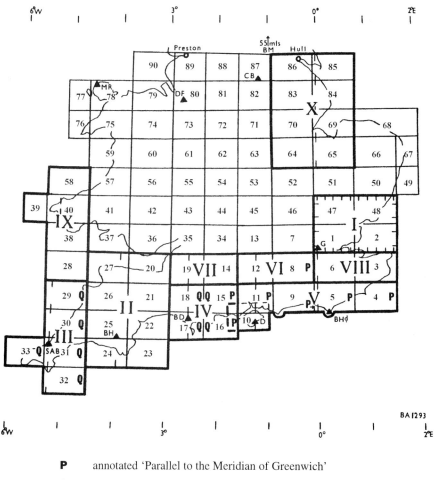

BA 1293

P	annotated 'Parallel to the Meridian of Greenwich'
Q	annotated 'Parallel to the Meridian of Butterton Hill'
▲	origin for the calculation of triangulation (list in text)
⊥	meridian through the above (tick inside rectangle indicates a line across and/or outside border on the actual map)

Preliminary investigations into the projections of the Old Series one-inch maps

The few mathematical cartographers, myself included, who have attempted to evaluate the above projections have been daunted by the sheer magnitude of the investigations which would be required. But to provide our chairman, Yolande Hodson, with some background to the projection used for the Popular Edition maps, which would assist her research into that series, I did some preliminary work on the Old Series which brought to light inter alia the startling facts about Butterton Hill meridian recorded above. To avoid overloading the main article with the more technical aspects of the investigations, I separated them out and discuss them here.

1. The calculation and setting down of the early surveys

It was realised in the early years of the Ordnance Trigonometrical Survey that not enough was then known about the size and shape of the Earth to carry out a rigorous adjustment of the main triangulation, but it was still necessary to calculate the positions of the triangulation stations in terms of rectangular co-ordinates so that they could be plotted on the copper plates of the new maps. At that time, too, the calculations were carried out by first adjusting the angles of each triangle to total two right angles, and then doing the computing as though the area concerned was flat.

William Mudge decided, very probably in consultation with Isaac Dalby, that the best way to take account of these factors was to divide the survey into several portions, in each of which one triangulation station was selected as an origin to which the rest were referred in terms of feet from the meridian and from the perpendicular; in other words in feet east or west and north or south. Mudge initially chose six origins spread across southern England at intervals of about sixty miles: Greenwich, Beachy Head, Dunnose, Black Down, Butterton Hill and St Agnes Beacon, although Beachy Head was only used for a comparatively small area. As the survey proceeded northwards four more origins were selected at Clifton Beacon, Delamere Forest, Moel Rhyddlad and Burleigh Moor (from the second of which 'Forest' was conventionally dropped by mid-century). The principal stations having been computed, co-ordinates for the secondary stations were calculated on the same origins, and the detail surveys were then fitted in, making the way clear for the one-inch mapping to be plotted from those origins.

But, referring specifically to southern England, the origins other than Greenwich were not used in constructing the Old Series sheet lines, and hence, though without any surviving evidence of the actual draughting of those sheets, it is evident that the positions of the origins and the directions of their meridians had first to be set down within the sheet lines before the rest of the detail could be drawn in relation to those origins. So we find the lines of these meridians engraved in the borders and/or margins of the southern swathe of Old Series maps, Greenwich on 5 and 6, and 1 and 47 (of which more below), plus 65, 69, 84, 85 farther north, Dunnose on 10, 11, 12, Black Down on 17, 18, 19, Butterton Hill on 24 and 27, and St Agnes Beacon on 32, 31, 30 and 29. If only this practice had been continued, investigation into the Old Series projections could have been much facilitated.

2. The graduated border round Part I

The first Old Series sheets showed evidence of transition between the privately produced eighteenth century county maps and the scientifically based 'General Survey of England &

Wales'. Initially they mapped particular counties, but were already distinctive in being completed to the neat lines across adjoining counties; in addition Part I was provided with the then familiar graduated border, not constructed in conformity with the surveyed detail on the map, but quite probably drawn round the four-sheet map after completion. Technically, the graduation was constructed on the Modified Plate Carrée Projection, on which the parallels of latitude were equally spaced horizontal lines and the meridians of longitude were equally spaced vertical lines.

Referring to the draughting methods outlined in section 1, Greenwich Observatory was plotted in its correct latitude and the meridian of Greenwich ran correctly through it, but the further one departed from the zero meridian, the greater the discrepancies between the notional vertical meridians and their true paths across the mapped detail, and between the notional straight parallels and their true curved equivalents. The somewhat obscure note[1] below the south-west corner of sheet 1 was presumably intended to warn the map user of these points, but I advance it as a purely personal suggestion that it may also show evidence of an early Ordnance Survey controversy as to the propriety of the inclusion of a noticeably inaccurate border on a properly surveyed map of the new national survey.

3. The sheet lines of the earlier Old Series sheets

As I say in my opening paragraph, the sheet lines of Part II were parallel and perpendicular to an unspecified meridian; but the longitude of that meridian can be deduced from the angle between the sheet lines and the marked meridian of Butterton Hill, known as the convergence of those meridians. It must be appreciated that precise figures cannot be obtained from measurements made on a mounted copper proof, which has shrunk in the proving process and been distorted in the mounting process, but figures accurate to within ten minutes of orientation should be possible. As my measurements suggest that the sheet lines of Part II were parallel and perpendicular to the meridian of 3° 04' West, there seems little doubt that the chosen meridian was, in fact, 3°W.

The sheet lines of Part III were parallel to those of Part II, and were therefore also parallel and perpendicular to the line of 3°W, and I remark here that it was the fact that the meridian of St Agnes Beacon, as indicated by the markers in the margins, appeared to run across Part III at much too great an angle that first alerted me to the fact that the sides were not parallel to the meridian of Butterton Hill as they were labelled.

Returning to the eastern side of the country, as well as Part I, the sheet lines of Parts V (excluding sheet 10), VI, VIII, and X farther north, were all firmly based on the Greenwich meridian, and it is in the area between the sheet lines parallel to 0° and those parallel to 3°W that the real complications start. Not only were the bounding sheet lines of this area converging towards each other northwards, but their distance apart was nowhere near a multiple of a standard sized sheet. What the thinking ahead had been when Part II was schemed can hardly even be guessed at, except that longitude 3°W must have been seen as a reasonable central meridian for the western part of Britain as a whole, for it ran quite near the edge of Part II and 4°W would have been more central to that part if a round degree had been thought desirable for its own central meridian.

---

[1] 'The Scale of Latitude and also that of Longitude around this Map, being drawn & graduated on a plane projection, the Latitudes & Longitudes deduced therefrom, can only be nearly true near to the Meridian of Greenwich' – see page 46.

Dealing first with the particular case of sheet 10, the Isle of Wight, although belonging to Part V it was treated very much as a special sheet and was displaced from the basic sheet pattern. Well known to be the first map to be labelled 'Ordnance Survey', my measurements show its central meridian to have been remarkably 1° EAST. This choice must surely have been made to dispose the island more neatly within the sheet lines, a course to which I for one take no exception. Proceeding north to sheets 11 and 12, I have already indicated that they were constructed on the Greenwich meridian and they were rectangular, as sheet 13 was also; they were, however, narrow, the space between sheets 9 and 21 having been divided into three to accommodate sheets 11, 15 and 18, through which two tapering columns of sheets rose from 17 to 89 and from 16 to 88, whilst the third only tapered from 45 to 87.

A short cogitation on those facts will indicate that some very peculiarly shaped sheets could be found in this central trapezium, compounded by the fact that additional origins were brought into use in its northern part. It is in this area in particular that much more work, including some very detailed measurements, needs to be done before any conclusions can be reached on the projections, or possibly lack of projections, involved.

Sheetlines 38, January 1994

Parallel to the Meridian of Butterton Hill – a correction

A regrettable error crept into my piece '"Parallel to the Meridian of Butterton Hill" – do I laugh or cry?' which appeared in *Sheetlines 38*. In the penultimate paragraph I stated that (Old Series) sheet 13 was 'also rectangular', whereas in fact its north border slopes downwards to the east at an angle of 0° 51' to the horizontal. The important thing in the context was that the main body of the sheet is rectangular so that the western and eastern borders are parallel, and, as stated, the third column of narrow sheets only tapers from 45 to 87. However, the reference in *Sheetlines 38* as it stands is erroneous, it should not have occurred, and I apologise for it.

The angle of 0° 51' may sound very small, thinking of a school protractor, but it does mean that the eastern border of the sheet is shorter than the western by 0.43 inches (1.1 cm). The sloping border must have been incorporated to accommodate the tapering column of sheets to the north, and therefore indicates that sheet 13 was not schemed until the shape of things to come had at least been provisionally determined. This is just one more lead towards solving the puzzle of the compilation story of the central Old Series sheets.

Sheetlines 50, December 1997

198 years and 153 meridians, 152 defunct

"Most references to county meridians in OS publications for at least 60 years have contained errors." "Ordnance Survey's most inaccurate maps." "My only criticism of Guy Messenger's book." These are all matters that I have felt moved to write to *Sheetlines* about in the past year. But they have a common component, Cassini's Projection, and the Editor has suggested that some elucidation of this projection may be desirable for our less technical members. I am therefore putting these items together in a short series. It is hoped that our professional members will bear with the introduction to Cassini's Projection, but the Editor and I both feel it will be welcomed by the wider circle of those who are fascinated by maps, but are not so familiar with the ways they are brought into being.

Part One – Winterbotham, Cassini and Messenger

150 years and 150 meridians was the title of an article contributed to the *Empire Survey Review* by Brigadier H St J L Winterbotham in 1938.[1] The author confessed in his opening paragraph that his title was not a precise statement but an averaged one – in its then 147 years the Ordnance Survey had employed 153 fully authentic meridians in its mapping of Great Britain and the Isle of Man; the former DG excluded the numerous Irish meridians from his arithmetical survey although all had been established long before 1921. Since 1938 no new meridian has been adopted, but several had already been superseded by the time of the Survey's centenary and many more had gone or were obsolescent by the time of its sesquicentenary. By that time too the die had been cast which was to lead to the one single meridian remaining in use today, longitude 2°W, the central meridian of the National Projection and National Grid, a Transverse Mercator Projection on which much has been published in past decades.

Only one other of the 153 was a round figure; longitude 4°W was the central meridian for the first three series of the one-inch maps of Scotland which were drawn on Bonne's Projection, as were smaller scale maps of Scotland of the same period. (Bonne's is an equal-area pseudo-conical projection on which the scale is preserved along all parallels of latitude but on only the central meridian.) The remaining 151 were central meridians of Cassini projections, two of national extent and many countywide, but the majority serving only the old large scale surveys of single towns. These meridians were drawn through the origins of the projections, which were normally specific topographical features upon which trigonometric stations were sited, and whose geographical positions were calculated through the national survey. Very few of the town surveys were in a position to adopt primary triangulation stations as their origins, but somewhat surprisingly less than half of the county origins were primary stations.

Cassini's Projection was a very basic one in which distances on the ground were plotted true to scale along the central meridian (CM) itself and at right angles to it, along great circles. If one imagines a globe turned on its side with its equator lying along the CM, it can be seen that projection distances parallel to the CM are increasingly greater than true distances on the globe as one moves away from that CM; this distortion is of no noticeable extent in a local survey, and of manageable proportions across the width of a county, but becomes a considerable nuisance over a wider area, where its problems are more

---

[1] Winterbotham, H St J L, '150 years and 150 meridians', *Empire Survey Review*, 4 (1938), 322-326.

conveniently coped with by a more sophisticated projection such as the Transverse Mercator. A short technical description of Cassini's Projection and tables for its construction were included in *Methods and processes*.[2]

What is not always appreciated by the layman is that in constructing most maps the familiar latitudes and longitudes, properly referred to as geographical co-ordinates, are no use in themselves and have to be converted to x and y rectangular co-ordinates by the projection formulae to enable the map detail to be drawn in. In modern times the projection co-ordinates have a dual purpose, serving also as the basis of a reference system; x and y become easting and northing. Co-ordinate lines are drawn as a grid on the face of the map, and usually form its sheet lines. But this is nothing new; all the sheet lines of the New Series, Third and Popular Edition one-inch maps of England and the Popular of Scotland are actual co-ordinate lines of a national Cassini projection, the obvious difference from today being that they are not numbered. So, too, the system of two-inch squares drawn across the face of a Popular Edition map is in modern terminology an un-numbered grid.

The reason the lines are not numbered will become apparent if we look for example at the fenland city of Ely, situated in square E9 on Popular Edition sheet 75; the co-ordinate lines bounding this square are:-

278,420 feet south

651,030 feet east　　　　　　　　661,590 feet east

288,980 feet south

a far cry from the round thousands of figures of the National Grid squares. The origin of this nationwide Cassini Projection was an early primary station in Cheshire, Delamere, situated at an internal point on Popular Edition sheet 44, the unit was the foot (the Ordnance Survey's own 'foot of 0_1'), and the co-ordinates were measured in four directions therefrom. Delamere was adopted as the origin of the projection and sheet lines of the Old Series maps from the Preston-Hull line northwards and then for their extension south as the New Series, and their conversion in turn to the Third Edition, small and large sheets, and the Popular Edition, and in due course for the extension of the latter edition to Scotland. It was also used for all the smaller scale maps derived from those series.

The position of Delamere was

latitude　　　　53° 13' 17".274 N

longitude　　　2° 41' 03".562 W

and all the sheet lines of the map series mentioned above are parallel or perpendicular to the latter meridian. The co-ordinate lines bounding Popular sheet 44 are:-

48,940 feet north

3,690 feet west　　　　　　　138,870 feet east

46,100 feet south

and all standard size sheets are 27 miles × 18 miles, or 142,560 feet × 95,040 feet. Hence, using a sheet index, the interested member can amuse him/herself by computing the Cassini co-ordinates on the Delamere origin of, say, his/her local Popular Edition sheet lines and thence his/her home; the diagram illustrates this exercise in relation to the Ely square mentioned above. This type of operation, formerly particularly familiar to the professional

---

[2] Ordnance Survey, *Account of the methods and processes adopted for the production of the maps of the Ordnance Survey of the United Kingdom*, second edition, London: HMSO, 1902.

surveyor using the county series plans, is equally applicable to the one-inch and smaller scales, except that direct reference to the maps may be necessary to carry the arithmetic through where non-standard or overlapping sheets are involved. Caution is also required in extending this exercise to Scotland through the common sheets 3 & 86 or 5 & 89, because of the one-mile overlaps introduced on the Scottish Popular Edition as well as its many non-standard sized sheets.

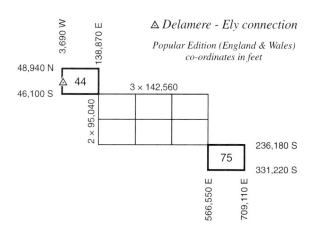

$\triangle$ *Delamere - Ely connection*

Popular Edition (England & Wales)
co-ordinates in feet

Sheet 75

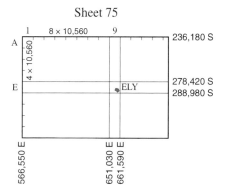

To complete the picture I give the data for the sheets containing Delamere in the one-inch Third Edition; the co-ordinates are:-

Small sheet 109

17,260 feet north

51,210 feet west   43,830 feet east

46,100 feet south

Large sheet 44

48,940 feet north

51,210 feet west   91,350 feet east

46,100 feet south

Third Edition small sheets are all 18 miles × 12 miles, or 95,040 feet × 63,360 feet, whilst the standard size large sheets have the same dimensions as the standard Popular Edition sheets. Hence we may again trace local Cassini co-ordinates on Delamere through the sheet line systems.

This leads on to the Society's *The Ordnance Survey ... Third Edition (Large Sheet Series)* by Guy Messenger;[3] I have but one criticism of this major monograph, and I may say at once that Guy agreed with it fully as soon as I mentioned it to him. It is the absence of any explanation in the opening pages of the items in the Sheet Histories headed 'Coordinates'. It is implicit in what I have said above that the columns of sheets in this series from nos. 3, 5 and 6 through 44 to 142 and 143 run effectively due north-south, but away from the Delamere meridian the lines of latitude and longitude run at a small angle to the sheet lines, gradually increasing to one of 3½° in sheets 68 and 88. This results in a difference of longitude of 2' 25" between the north-east and south-east corners of these vertical sheets.

Thus, strictly speaking, Guy Messenger's 'Coordinates' are the mean values of latitude and longitude along each of the four sides of the respective sheets, but they are in any case approximate, being quoted to the nearest minute only, and are provided basically for location purposes in the absence of any figures on the maps themselves. To be pedantic, it would have been better had they been headed 'Geographicals', for this is the conventional abbreviation for 'geographical co-ordinates' among professionals, whilst 'co-ordinates' unqualified would usually refer to rectangular projection co-ordinates. But this is clearly not a serious criticism, and I can but add at this late stage my congratulations and thanks to Guy Messenger for his monumental work.

Addendum – Towards a National Grid

The choice of a grid for British maps,[4] reporting a discussion meeting of the Royal Geographical Society in March 1924, records an early use, possibly the first public use, of the expression 'national grid'. But the first stirrings towards a national Ordnance Survey Grid, as distinct from a military grid, had appeared a few years earlier. The one-inch Third Edition (Large Sheet Series) had carried an alpha-numeric reference system marked off by cross-ticks in the borders at two inch, or more correctly two scale mile, intervals. Guy Messenger records that on four of the special district sheets of this series the ticks were joined across the face of the map forming a system of two mile squares. Starting from the north-west corner these were lettered from A southwards (omitting I) and numbered from 1 eastwards, finishing where necessary with intervals of odd lengths adjoining the east and south borders.

This system was then adopted for the Popular Edition maps of England & Wales, forming, as I remarked above, an un-numbered Cassini grid, but one which was individual to the sheet concerned. However, on sheets published from 1920 onwards the system was

---

[3] Messenger, Guy, *The Ordnance Survey one-inch map of England and Wales Third Edition (Large Sheet Series)*, London: Charles Close Society, 1988, ISBN 1 870598 03 2.

[4] Winterbotham, H St J L, 'The choice of a grid for British maps', *Geographical Journal,* 63 (1924), 491-503.

amended so that the squares formed part of a continuous semi-national two mile grid. Thus standard sheets 35 and 37 conformed to the pattern stated above with one mile wide rectangles numbered 14 at their east ends, but the adjoining sheets 36 and 38 had one mile rectangles numbered 1 at their west ends forming two mile squares with the aforementioned number 14s. The standard sheets were all nine squares deep (A to J), and so the portrait sheets 1 and 2 had rows of one mile rectangles lettered A against their north borders so as to finish with two mile squares (lettered O) against their south borders adjoining the standard sheets 3 and 4. The portrait sheets 8 and 18 had rows of one mile rectangles on both their west and east sides, numbered 1 and 10 respectively, in order to conform to the continuous grid line system.

This semi-national two mile grid was used for sheets 43 to 47, then 35 to 40 and all sheets northward except 17 (Isle of Man), and on the extension of the Popular Edition to Scotland was duly completed across the whole of that country. All these grid lines are co-ordinate lines of the national Cassini Projection on the origin of Delamere, and their co-ordinates are all multiples of 10,560 feet added to the particular co-ordinate of the relevant side of the two mile square containing Delamere; the co-ordinates of the sides of this square (which lies half on sheet 43 and half on sheet 44) are:-

6,700 feet north

8,970 feet west   1,590 feet east

3,860 feet south

Part Two – Confusion worse confounded, the gridded grid

Part One focussed largely on the application of Cassini's Projection to the one-inch map, the favourite of the purchasing public, as Brigadier Winterbotham termed them, but by whom the properties or even the existence of the projection were frequently unappreciated; subsequent parts will feature aspects of the six-inch and larger scale county series maps, wherein the projection often has a greater day-to-day impact on the user who is generally more familiar with its technicalities.

But here is someone else with a furrowed brow; he is, it seems, a Scottish ex-soldier. "You have explained" he says "how the Popular Editions of Scotland and England were constructed on a single Cassini Projection, with the sheet lines parallel and perpendicular to the meridian of Delamere. But I defended these lands using these maps printed with a Cassini grid which lay at an angle to the sheet lines, and then after the war the Scottish Popular Edition was reprinted with the National Grid, a Transverse Mercator grid running at another angle to the sheet lines. Can you explain these, please?" Indeed I will try.

My sub-heading may appear over-dramatic, for one grid would not normally be printed over another one, except under the exigencies of wartime operations. But as I have indicated in Part One, grid lines on a basic map are in fact projection co-ordinate lines and are therefore still present as an invisible net even if not ruled on the map. So, to take the example of the Scottish Popular Edition with National Grid, we (apparently) have a Transverse Mercator grid on top of a transparent Cassini grid; not wholly transparent because the sheet lines are Cassini co-ordinate lines. The vital point to grasp is that the map and everything on it are constructed on Cassini's Projection, and the National Grid is present here solely as a reference system. It is perhaps best in this instance to think of the grid lines as lines on the ground, cutting the topographical features exactly as they do on the Seventh Series map; the lines are then mapped on the Scottish Popular by Cassini's Projection along with all other ground detail.

Thus, whilst on the National Projection maps of today the National Grid is an exact system of squares forming both a construction network and a reference system, when printed on any other map it will appear only as a reference grid composed of curvilinear rectangles, although often (as in the present example) they are only very slightly distorted squares. A similar effect occurs on GSGS 3907 and 3908, the military editions of the one-inch recently discussed by Richard Oliver.[5] As stated thereon these maps carry the War Office Cassini Grid whose origin was Dunnose, a primary triangulation station in the south-east of the Isle of Wight and county origin for fifteen central counties. So even this grid, although a Cassini grid, was not quite an exact square grid when drawn on a map on Cassini's Projection with a different central meridian, i.e. Delamere; and with a different central meridian the grid lines lie at an angle to the map sheet lines, technically the convergence of the two meridians. Yet another variation appears on the eleven south-western sheets of the War Revision versions of GSGS 3907, which were Fifth Edition sheets and carried the War Office Cassini Grid over the Fifth Edition Transverse Mercator Projection, central meridian 2°W.

The reasons for the choice of Dunnose as the origin of the War Office Cassini Grid have never been published but are clearly connected with its prior status as the county origin for the central counties, especially Hampshire and Wiltshire which included many military establishments for which Dunnose co-ordinates would already have been in existence. The position of Dunnose is latitude 50° 37' 03".748 North, longitude 1° 11' 50".136 West, but the grid has a false origin 500,000 metres west and 100,000 metres south of Dunnose, and co-ordinates thereon are known as WOFO (War Office False Origin) co-ordinates; the grid itself was often familiarly referred to as the WOFO ('Woffo') grid.

To sum up, any grid may appear on any map for reference purposes, the projection of the base map remains paramount, and the quoted projection of the grid only refers in this case to the way it is laid down on the ground and not to its projection on the base map. The way it is laid down on that base map may confidently be left to the mathematical geodesist.

Terminology (applicable to Parts Three and Four)

When referring to the situation in the final stage of county series mapping the present tense is generally used.

'County' means a geographical county as existing prior to 1965, except that in the context of county series, Isle of Lewis, Isle of Skye and Outer Hebrides are equivalent to counties, having separate series of sheet numbers (Outer Hebrides here signifies just the Inverness-shire portion); the Isle of Man, too, is here equivalent to a county.

'Combined counties' means a group of counties, not necessarily contiguous, mapped on the same origin with a single system of sheet lines, though each county usually has its own series of sheet numbers.

'Unified counties' means a pair of combined counties with a single unified series of sheet numbers.

---

[5] Oliver, Richard, 'One-inch military maps of Great Britain, 1919-1950: some notes on GSGS 3907 and 3908', *Sheetlines 20* (December 1987), 9-12.

Part Three – Ordnance Survey's most inaccurate maps (?)

In 1920 the OS published a booklet *A description of the Ordnance Survey large scale maps*;[6] running through eight editions, the last published in 1954, it had a total shelf life of some forty years. Most editions contained two maps showing the origins used for the projection co-ordinates of the county series of twentyfive-inch and six-inch maps, which were drawn on Cassini's Projection. Covering respectively England & Wales and Scotland, they depicted the county origins and their meridians in red over black base-maps showing the counties. The same maps were included in Brigadier Winterbotham's *The national plans* published in 1934.[7] Throughout they were entitled 'Diagram ... showing meridians', the Ordnance Survey's usual practice being to describe a map as drawn on the meridian of so-and-so. But every point on a flat surface has to be fixed by two co-ordinates and it is their origin which is the definitive feature. Further, a plot of positions on a geographically true base-map, especially when produced by the OS, would hardly be regarded as a diagram.

However, most people will know that St Paul's is situated in the centre of London and not in the vicinity of Harrow in Middlesex, where it is marked on the England & Wales map; many will know that Leith Hill is as near the southern boundary of Surrey as it is depicted near the northern. Anyone familiar, if only as a map user, with the peaks which comprise the origins in the Scottish Highlands will realise that several of them are shown miles out of position; indeed Ben Cleugh placed in the middle of Perthshire actually lies in Clackmannanshire. A detailed check, inspired by these evident errors, on all thirty-eight origins suggests that they were thrown at the map from some way off, for their positions are virtually all in error by distance rising to 17 miles, the average error on the Scottish map being 5.6 miles and on the southern map 5.2 miles. A glance at the maps accompanying this article will show that the erroneous positions lie in all directions from their true positions, so that the meridians are as much in error as the origins themselves.

The red plates also carry firm lines enclosing the groups of combined counties, as do black ones on my maps, but in some editions of the OS booklet as well as *The national plans* West Lothian is incorrectly included with the counties drawn on The Buck instead of those on Lanark Church Spire. Several incorrect spellings of the names of the origins also persist through some or all editions of the maps. All these facts lead to the suggestion that these two maps must be among the OS's most inaccurate, and I find it surprising that such maps, produced for the information of map users and surveyors by the country's national cartographic authority, continued to be reprinted without major correction for so long. I find it equally surprising that Winterbotham, who one supposes was particularly familiar with the subject, should have included the maps in his Professional Paper without apparently realising the errors they contained, especially the incorrect assignment of West Lothian to the origin of The Buck.

For some reason the four south-western origins in England were not plotted on the maps in the first three editions of the OS booklet, being indicated solely by their names and meridians. In the original edition only, the remaining English and Welsh origins were marked by open circles, and Danbury Church Spire which (as will be explained in Part Four) lay

---

[6] *A description of the Ordnance Survey large scale maps,* OSO Southampton, undated (1920); second edition (1922); third edition (1926); fourth edition (1930); fifth edition (1937); sixth edition (1939). *A description of the Ordnance Survey large scale maps,* OS Chessington, unnumbered edition, 1947. *A description of the Ordnance Survey large scale plans,* OS Chessington, unnumbered edition, 1954.

[7] Winterbotham, H St J L, *The national plans*, Ordnance Survey Professional Papers, New Series 16, HMSO 1934.

outside the combined counties drawn on it, continued to be so marked throughout the life of the map. Otherwise, and on the Scottish map from the start, the origins were indicated by solid dots. There is no official list of the county origins and some of their names will be found in varying forms in OS internal documents and publications. I have examined as many of these sources as possible and drawn up a list of generally accepted forms as given on the page of abbreviations accompanying my two maps. It should be remembered that the names of the origins date from the early days of the Survey, and that the spellings adopted for them at that time remain unchanged in their roles as origins, notwithstanding any changes in their spellings as features. In addition to the definite spelling errors already mentioned, the maps in the OS booklet have some names differing in form from the approved norms, e.g. Brandon instead of Brandon Down.

The red plates were re-prepared for the fourth (1930) edition and it was at this stage that the combined counties' boundary was diverted to the wrong (west) side of West Lothian; it was corrected to run east of this county in the fifth (1937) edition. The sixth (1939) edition contained a single newly-drawn map of Great Britain in black only, but it must be presumed that this was lost in the bombing as the first post-war edition of 1947 reverted to the old pair of two-colour maps and to the erroneous assignment of West Lothian! The latter was re-corrected in the final 1954 edition, which also for the first time correctly spelt Scour-na-Lapich (previously -Lopach) though it included the hyphens not generally used. For this last edition the title of the booklet was altered to *A description of the Ordnance Survey large scale plans*.

Even the black base-maps (excluding the sixth edition) contained some strange features; the old county names of Carnarvon, Edinburgh, Elgin, Forfar, Haddington and Linlithgow appearing in the first edition were all altered to their modern forms in the fourth, but the 'm' in Dumbarton was never corrected to 'n', and the older alternative spelling Argyle was used throughout although this spelling was not otherwise used by the OS as far as I can determine. These two counties were spelt correctly in the sixth edition. But the oddest feature of the base-maps was the remarkably idiosyncratic selection of towns; only one (Aberdeen) appeared on the Scotland map, and none appeared in thirty-six counties of the southern map, whereas, for example, Lincolnshire contained Grimsby and Grantham, and in the East Riding the village of Flamborough was marked together with Hull. In the post-war editions the positional circle at Flamborough was deleted, but although the name might have been regarded as applying to the promontory the word 'Head' was not added; in these editions too an intrusive 'o' appeared in Middlesboro'.

In 1975 a new publication appeared, *Ordnance Survey maps, a descriptive manual* by J B Harley,[8] replacing the former three booklets on different scales. Unfortunately it includes a single plate of Great Britain combining the material on the pair of maps in the former large scale booklet; the glaring errors in the positions of the origins and meridians (now on a blue plate) are all reproduced, as well as the peculiar selection of towns plus the name Flamborough. Three spellings are corrected and the abbreviation for 'Island' altered, but Argyle and Dumbarton remain, and an additional error is introduced in the boundary of Peebles on the black but not on the blue. These matters can hardly be laid at Brian Harley's door, for the plate is clearly the result of an unthinking exercise by the drawing office in combining the two old maps, but it has to be regarded as a prime contender for the Ordnance Survey's 'most inaccurate map' title.

---

[8] J B Harley, *Ordnance Survey maps a descriptive manual*, Southampton: Ordnance Survey, 1975.

I cannot forbear to add that, apart from a small displacement of Lanark, none of the catalogue of errors appears on the maps in a booklet issued by the Hydrographic Department for its surveyors and staff in 1946 (in which a young hydrographic officer named B W Adams had a considerable hand).[9] On another page this booklet does contain incorrect co-ordinates for the sheet lines of Zetland 54 (Foula), but my research into the present series has confirmed that the erroneous figures were supplied from, and still remain in, the OS records at Southampton and Kew.

Lanark Church Spire – but which?

Brian went to Lanark in 1995 to investigate the uncertainties surrounding the actual location of the trig. station Lanark Church Spire, generally assumed locally to be the parish church, St Nicholas (B on the plan he drew following his visit). His findings are summarised here.

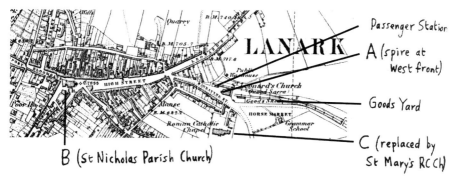

Brian's first qualm on approaching Lanark from the east was to find the skyline dominated by a spire (C) southward of the expected place; this turned out to be one which had replaced a burnt-down predecessor which itself had only been in the course of construction at the time of the first OS large scale survey, and had therefore not been used as a trig. station. But his main surprise was to find the spot where he had expected to find another church, St Leonard's (A), occupied by a Job Centre. The church had been demolished in 1969.

He quoted the fundamental data for Lanarkshire from the Ordnance Survey sheet line co-ordinate data referring to the old County Series plans:

Origin: Lanark Church Spire, Lat. 55° 40' 24".170 N., Long. 3° 46' 18".370 W.

	18,120 feet north	
Sides of six-inch Lanarks sheet 25:	19,680 feet west	12,000 feet east
	3,000 feet south	

By transferring these figures to Lanarks 1:500 sheet XXV.15.13, he confirmed that the origin of the county co-ordinate system is the trig. point shown in the position of St Leonard's church spire (A), the trig. point being marked at 672 scale feet from the western neat line and 168 scale feet from the northern neat line of the 1:500 plan.

[9] HD 364, *Ordnance Survey sheet line co-ordinate data for six inch and 1:2,500 sheets*, Hydrographic Department of the Admiralty, Restricted Issue 1946.

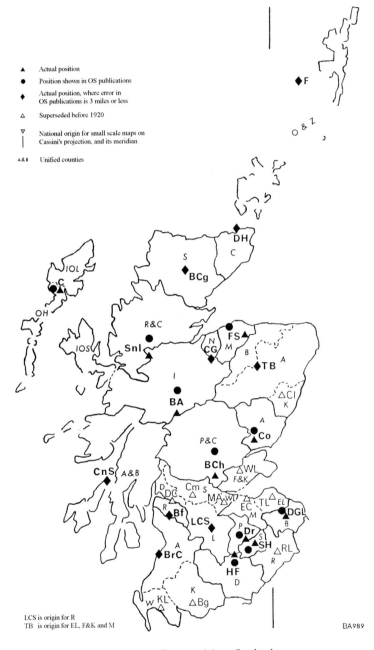

County origins – Scotland

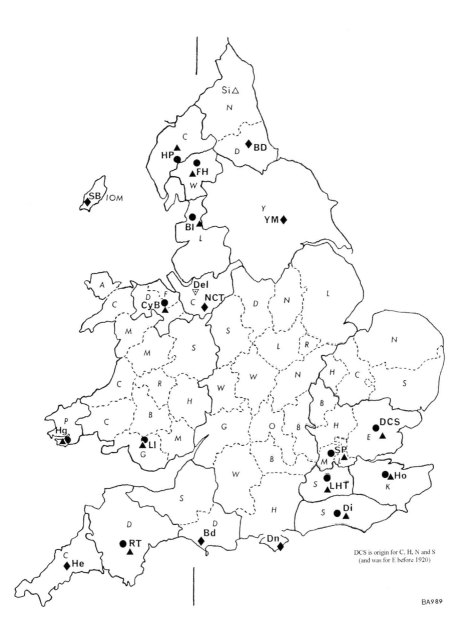

DCS is origin for C, H, N and S
(and was for E before 1920)

BA989

County origins – England and Wales

List of county origins

Code	Name of origin	Latest form		Geographical position		Identification on 1:50,000 map		
				latitude N	longitude	N G ref	m.	Name at point
BA	Ben Auler	Ben Alder		56° 48' 50'.431	4° 27' 49'.113 W	NN 496719	1148	
BCh	Ben Cleugh	Ben Cleuch	P	56 11 08.695	3 46 03.873 W	NN 903006	721	
BCg	Ben Clibrig	Ben Kilbreck	P	58 14 07.844	4 24 35.534 W	NC 585299	961	Meall nan Con
Bg	Bengray			54 54 51.940	4 08 11.178 W	NX 631598	366	
Bd	Blackdown	Black Down	P	50 41 10.287	2 32 52.432 W	SY 613876	237	Hardy Monument
Bl	Bleasdale [a]			53 54 55.388	2 37 20.809 W	SD 591468	510	Fair Snape Fell
BD	Brandon Down		P	54 45 17.630	1 40 36.632 W	NZ 208401	266	
Bf	Broadfield			55 47 59.798	4 32 20.609 W	NS 409592	236	
BrC	Brown Carrick	Brown Carrick Hill		55 24 26.531	4 42 41.070 W	NS 284160	287	
Cl	Caerloch	Kerloch		56 58 52.543	2 29 56.890 W	NO 697879	534	
CG	Cairn Glasher	Carn Glas-choire		57 20 22.829	3 50 30.775 W	NH 891292	659	
C	Cleisham	Clisham	P [m]	57 57 50.561	6 48 41.648 W	NB 155073	799	
			[n]	57 57 47.350	6 48 42.073 W			
Cm	Corkmulaw	Cort-ma Law		55 59 36.745	4 09 44.286 W	NS 651799	531	
Co	Craigowl	Craigowl Hill		56 32 52.216	3 00 48.599 W	NO 377400	455	
CnS	Cruach-na-Sleagh	Cruach na Seilcheig		56 07 08.247	5 43 34.944 W	NR 684981	296	[b]
CyB	Cyrn-y-Brain	Cyrn y Brain		53 02 16.870	3 10 22.300 W	SJ 214496	562	
DCS	Danbury Church Spire		P	51 42 57.897	0 34 32.746 E	TL 779051		
Del	Delamere [c]		P	53 13 17.274	2 41 03.562 W	SJ 543696	176	Pale Heights
DGL	Derrington Great Law	Dirrington Great Law	P	55 47 12.070	2 28 51.930 W	NT 698549	398	
Di	Ditchling			50 54 04.039	0 06 21.838 W	TQ 332131	248	Ditchling Beacon
DC	Dumbarton Castle			55 56 12.478	4 33 46.136 W	NS 399745		
DH	Dunnet Head		P	58 40 09.962	3 22 13.056 W	ND 205765	127	
Dn	Dunnose		P	50 37 03.748	1 11 50.136 W	SZ 568802	235	Shanklin Down
Dr	Dunrich	Dun Rig	P	55 34 19.950	3 11 01.472 W	NT 254316	742	
EC	Edinburgh Castle			55 56 54.110	3 11 53.440 W	NT 252735		
FS	Findlay Seat	Findlay's Seat		57 34 42.276	3 14 26.599 W	NJ 258549	338	Brown Muir
FH	Forest Hill	The Forest		54 25 30.680	2 43 41.580 W	NY 528036		
F	Foula		P	60 08 26.609	2 05 38.668 W	HT 948395	418	The Sneug
HF	Hart Fell		P	55 24 28.961	3 24 00.212 W	NT 114136	808	
He	Hensbarrow	Hensbarrow Downs	P	50 22 58.810	4 49 05.137 W	SW 997575	312	
Hg	Highgate		P	51 39 37.750	4 56 17.995 W	SR 968999		

Code	Name of origin	Latest form		Geographical position latitude N	longitude	Identification on 1:50,000 map N G ref.	m.	Name at point
HP	High Pike			54 42 19.170	3 03 26.600 W	NY 319350	658	
Ho	Hollingbourne			51 16 10.380	0 39 55.564 E	TQ 859557	197	
KL	Knock of Luce			54 51 58.316	4 43 10.206 W	NX 255558	175	Knock Fell
LCS	Lanark Church Spire			55 40 24.170	3 46 18.370 W	NS 886436		[c]
LHT	Leith Hill Tower		P	51 10 32.887	0 22 11.049 W	TQ 139432	294	
Ll	Llangeinor	Mynydd Llangeinwyr		51 38 26.680	3 34 17.000 W	SS 913948	568	Werfa
MA	Mount Airy			55 55 22.038	3 37 05.376 W	NS 989711	312	
NCT	Nantwich Church Tower			53 04 00.152	2 31 09.186 W	SJ 652523		
OCT	Otley Church Tower		P	52 08 53.789	1 13 17.963 E	TM 204549		
RT	Rippon Tor [a]			50 33 57.034	3 46 12.078 W	SX 747756	473	
RL	Rubers Law			55 25 55.971	2 39 48.069 W	NT 580156	424	
SP	St Paul's Cathedral [d]		P	51 30 47.753	0 05 48.356 W	TQ 321811		
SH	Sandhope Heights	Sundhope Height		55 30 10.253	3 02 31.092 W	NT 342237	513	
Snl	Scournalapich	Sgurr na Lapaich	P	57 22 10.334	5 03 31.678 W	NH 161351	1150	
Si	Simonside [a]			55 16 50.750	1 57 26.410 W	NZ 027985	429	
SB	South Berule	South Barrule	P	54 08 57.597	4 40 05.328 W	SC 258759	483	
SC	Stafford Castle [o]		P	52 47 51.000	2 08 44.580 W	SJ 902223		
TB	The Buck			57 17 18.978	2 58 11.630 W	NJ 415224		Kebbuck Knowe
TL	Traprean Law	Traprain Law	[m]	55 57 48.110	2 40 13.380 W	NT 582747	221	
WL	West Lomond		[n]	56 14 43.980	3 17 43.583 W	NO 197066	522	
YM	York Minster		P	53 57 43.265	1 04 49.752 W	SE 603522		

[a] properly Bleasdale new, Rippon Tor new, Simonside new (names of trig. points)

[b] adjacent to Sir Watkin's Tower

[c] not on map: position of church is just within circle of railway station symbol

[d] conventionally St. Paul's

[e] Delamere is not a county, but a national origin
    accepted value

[m] value used originally for sheet graduations

[o] properly Stafford Castle (centre)

[P] principal triangulation station

No key is provided to the initials of counties; readers unfamiliar with the British counties prior to 1965 are referred to an appropriate map.

Part Four - The story of the county origins

Most of the textual references to county meridians or county origins in OS publications for at least sixty-five years have contained errors, particularly as to the numbers in use at different times, whilst their graphical depiction was shown in Part Three of this series to be a cartographic disaster area. Yet they were fundamental to much the greater part of OS mapping up to the Second World War. Further words on this wider aspect fail me, but it is the purpose of this present part to record the true numbers involved. *A history of the Ordnance Survey*, chapter 21,[10] mentions some of the differing arithmetic in previous publications and remarks 'These discrepancies are not easy to resolve ... it is difficult to establish the initial situation and the subsequent history.' Frankly, I find this nonsense; the bones of the story are adequately documented even though some of the flesh is missing, and a day-and-a-half at Southampton, two afternoons at Kew, and some hours at Parsons Green on my calculator, sufficed to elicit the relatively few details of the old county origins which I did not know already. Chapter 21 also wrongly assigns the transfer of West Lothian to The Buck, which can only have been suggested by the incorrect editions of the maps described in Part Three. I have to admit here that some of the faulty arithmetic on county meridians is in Brigadier Winterbotham's 1938 article,[11] whilst some of his other figures are rather suspect, and these facts cast doubt on the figures in my series title, though there is no question that they are all defunct but one.

The reasons for the adoption of county series mapping by the OS have been frequently aired and I do not wish to expand on them here; very briefly it was the normal practice at the time of the foundation of the Survey. County-based (large scale) OS mapping in Great Britain started in 1840 with the six-inch surveys of Lancashire and Yorkshire, using the origins of Bleasdale and York Minster; these were followed by seven Scottish counties (see Terminology, page 58), and immediately two anomalies appear. Cleisham, the origin selected for Isle of Lewis, lay three miles outside the county boundary. Although this was unusual for the time, Cleisham was a primary station of the old triangulation, conveniently situated to be on the edge of the Lewis sheet line system. Whether there was any thought that it would be used for the rest of the Outer Hebrides it is impossible to guess. Secondly, Fife and Kinross were published as unified counties on the origin of West Lomond. It seems to have been early recognised that because of the small size of a few Scottish counties, and the strange configurations of some prior to the extensive boundary rationalisations of 1889-91, the adoption of unified sheet line systems for some pairs of counties, one large and one small, was the only reasonable course. In such circumstances both counties were mapped together on the relevant sheets, whereas the early large scale plans were not otherwise filled to the borders with adjoining counties.

Returning to Fife and Kinross, Winterbotham (1938) makes the categorical statement 'Even Kinross-shire ... had its own meridian', but had it? The actual survey consisting of angles and chainings had no origin; only when it reached the computer (a human being armed with log tables) had meridian and origin to be selected for the calculation of co-ordinates. There were certainly instances when the origin chosen initially for these calculations was superseded by another before the publication of the relevant maps, and such may have occurred in Kinross. But there is no evidence of it at all in the surviving material I

[10] Seymour, W A (ed.), *A history of the Ordnance Survey*, Folkestone: Dawson, 1980.
[11] Winterbotham, H St J L, '150 years and 150 meridians', *Empire Survey Review*, 4 (1938), 322-326.

have examined at Southampton and Kew, and it seems just possible that Winterbotham, writing three years after his retirement, was relying on an imperfect memory. The other four counties first surveyed and published at the six-inch scale were Edinburgh (later Midlothian) on the origin of Edinburgh Castle, Haddington (East Lothian) on Traprean Law, Kirkcudbright on Bengray, and Wigtown on Knock of Luce.

Thereafter (glossing over the battle of the scales) the twentyfive-inch took over as the ruling scale, but the same pattern of county mapping continued, with Clackmannan & Perth and Argyll & Bute published as unified counties. The 'skeleton' survey of London, undertaken forty years before the creation of the County of London, adopted St Paul's Cathedral as its origin, and this was followed in 1862 by the full survey, which was in effect the most extensive of the old town surveys. The Middlesex county survey was executed concurrently with this and naturally used the same origin, but the Hertfordshire survey commenced three years later was also published on the meridian of St Paul's. Although Middlesex and Hertfordshire were drawn on the same origin with a single continuous system of sheet lines, they were not published as unified counties, as defined in my terminology, but formed two separate series of maps, each individually numbered. However the secondary and tertiary surveys were linked across the common boundary and the cross-boundary sheets were joint sheets with dual numbers in both series.

This first publication of combined non-unified counties was tangible recognition of the difficulty caused by adjoining areas having been independently surveyed and mapped and simply not joining properly. It set the pattern for the greater part of the remaining county series mapping with blocks of combined counties of increasing size, culminating in twelve Welsh and Marcher counties on the origin of Llangeinor, and fifteen central English counties on the origin of Dunnose. For some reason, however, combined counties found little early favour in Scotland apart from the unusual case of Aberdeen and Banff, two more pairs of unified counties (Ross & Cromarty – then still separate counties – on Scournalapich and Orkney & Shetland, or Zetland, on Foula), and the special cases of the Inner and Outer Hebrides. The origin selected for Aberdeen was The Buck, which was also well situated to be that for Banff although at that time it was four miles outside the latter county. So the two were mapped as combined counties but with the cross-boundary sheets issued as two individual sheets, neither filled to the border with the other's ground. Yet these counties were, and always remained, indexed together on the same sheet and I suspect they may have been intended originally to be unified counties, but were numbered separately under the influence of the almost simultaneous publication of Middlesex and Hertfordshire.

This phase of county mapping was completed in 1891, with all but the original nine six-inch counties having been published on both the twentyfive- and six-inch scales, and there was a total of forty-nine county origins in use in Great Britain and the Isle of Man. There followed 'one of the Ordnance Survey's worst errors' (*A history of the OS*), the infamous episode of the Replotted Counties whose twentyfive-inch maps were largely drawn from the old six-inch surveys from 1887 onwards. The only good thing to emerge from this operation was the tardy emergence of two sets of combined counties in Scotland, where the six mainland counties involved were transferred on to two other existing origins, Kirkcudbright and Wigtown being redrawn on the Ayrshire meridian of Brown Carrick, and Edinburgh, Fife & Kinross and Haddington being redrawn on The Buck. It is a pity that this aspect of the replotted counties operation is omitted from the Seymour-edited history as it was the first time that the plans of a published county were transferred, that is, redrawn on a new origin with a new set of sheet lines on a slightly different orientation. But an even more significant operation executed a few years later, over the turn of the century, was the transfer of Kincardine from its first origin of Caerloch also on to The Buck. This was the first transfer of a county which had already been published on the twentyfive-inch scale, predating the next by a dozen years, and yet it seems almost certain that it has never been mentioned in print. It was not described in OS reports of the time, nor in the more recent standard works, and I am indebted to Michael Wood and his colleagues at Aberdeen University for confirmation that it does not seem to have been recorded locally. A glance at the map will suggest that the transfer of the Forthside replotted counties on to The Buck must have been carried out with the eventual transfer of the intervening counties of Forfar (Angus) and Kincardine also in mind, yet the whole process of thought and partial(?) execution appears destined to remain an intriguing mystery.

There were now still forty-three origins in use and the problems of adjoining counties on different origins were becoming more pressing, especially where built-up or industrialised areas crossed the boundaries of non-combined counties. It should also be mentioned that most of the principal estuaries had their opposite shores in different counties, and the hydrographic surveyor, working for example in the Thames Estuary, could fix his position from up stations in Essex or from those in Kent, but he could not obtain a fix from a mix of stations on both sides of the estuary. So it was that Colonel Charles Close decided in 1912 to institute a rolling programme to transfer one-by-one as many counties as possible on to other existing origins, but still recognising that Cassini's Projection was not suitable for areas wider than about three average counties at the large scales. The procedure was firstly to compute the co-ordinates of the relevant primary and secondary stations of the new origin, then to re-compute the tertiary points from the old observations on to the new positions, and hence to redraw the plans on the new sheet lines. It has to be said, however, that combining independent surveys on to a single map does not eliminate the intrinsic problems, but merely presents them in a new form.

The programme, commenced in earnest in 1913, was interrupted by the First World War, and had to be abandoned 'temporarily' in the post-war stringencies of 1919, never to be resuscitated. By 1919 six counties had been transferred and five origins, not six as has often been stated, had been eliminated, leaving a total of thirty-eight in use for the final period of county series mapping. In Scotland, Dumbarton (later Dunbarton), Linlithgow (West Lothian), Roxburgh and Stirling were all transferred on to Lanark Church Spire, whilst in England, Northumberland was transferred on to Brandon Down and Essex on to St Paul's. This programme eliminated the old origins of the first five (Dumbarton Castle, Mount Airy, Rubers Law, Corkmulaw, Simonside, respectively) but the former Essex origin, Danbury Church Spire, still remained in service for Cambridge, Huntingdon, Norfolk and Suffolk, although outwith that combined area. When the transfer programme ceased the data had been re-computed for five more counties – Cambridge, Huntingdon and Suffolk to go on to St Paul's, Dumfries and Selkirk on to Lanark Church Spire – and (not previously recorded) the new sheet line patterns had been devised and the latitudes and longitudes of the full six-inch sheet corners calculated for four more – Norfolk (sheet numbering provisional), Berwick, Edinburgh (Midlothian) and Peebles, on to the same two origins.

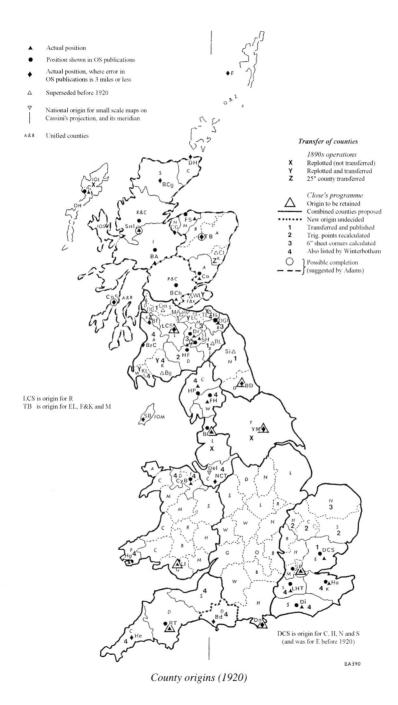

County origins (1920)

BA390

Winterbotham (*The national plans*) states in his Chronology, section 27, but not in the main text of section 12, that Close intended to reduce the number of meridians to eleven. He also lists in section 12 what he terms 'the remainder of Close's programme' but which only includes the counties south of the River and Firth of Forth, and shows that these would have been combined into eight blocks. The limits of these blocks, the origins involved, and the stages reached in the different counties are illustrated on the map herewith; it will be noticed that four of the replotted counties were to have been transferred a second time![12] Looking at the areas north of the Forth and west of Loch Long, I conclude that Close's supposed target of eleven could only be met if his programme were restricted to the mainland of Great Britain plus the Inner Hebrides, and my map also depicts a purely personal suggestion as to how the programme might have been completed on that basis.

I finish with a table summarising the numbers and changes of county origins in Great Britain and the Isle of Man:-

	England & Wales	Isle of Man	Scotland	Total
Origins for partial series (25-inch only) transferred c.1880 on to two of below	2			2
Origins used for the first county plans either six-inch or 25-inch, excluding partial series	19	1	29	49
Replotted counties, total	2		7	9
(Fife, Kinross counted separately)				
transferred (ditto)			6	6
origins eliminated			5	5
Mystery transfer of Kincardineshire (see page 89)			1	1
Origins in use 1902-1913	19	1	23	43
Close's programme, counties transferred	2		4	6
origins eliminated	1		4	5
Origins remaining in use 1919-1924[13]	18	1	19	38
Total all time origins	21	1	29	51

Sheetlines 25, August 1989; Sheetlines 26, December 1989; Sheetlines 27, April 1990 plus other unpublished material. Part Five of the original series will be found on page 79.

Part Five of the original series will be found on page 79.

---

[12] Winterbotham, H St J L, *The national plans*, Ordnance Survey Professional Papers, New Series 16, HMSO 1934. *The replotted counties*, supplement to the above, official use only, OSO Southampton, 1934.

[13] 1924 – withdrawal of last Old Series map.

Ordnance Survey County Series plans – sheet line data
Great Britain and Isle of Man

This compilation is tailored for use with individual county sheet indexes. Readers using Ordnance Survey book indexes or reprints thereof are warned that not all relevant sheet numbers are shown thereon, and certain opposite pages do not align. Background information on the series can be found in the compiler's introduction to David Archer's 1993 reprint of the Ordnance Survey Indexes for Scotland circa 1907, reproduced at page 84.

Data sources:

24% of items are from old Ordnance Survey listings.

19% are from the compiler's extensive research in the Public Record Office, Ordnance Survey Archives, Military Survey, and British Library Map Library. It is this invaluable and unique material which makes the data complete.

57% are compiler's computations.

Notes:

'-shire' is routinely omitted where separable.

Names of county origins date from the early days of the Survey and their spellings from that time remain unchanged in their roles as origins, notwithstanding any changes in their spellings as features.

Some names, especially in Scotland, are quite misleading as to their actual sites.

Defining elements of County Series systems:

Projection	Cassini's	
Spheroid	Airy	
Horizontal Datum	OSGBI 1858	
Unit	foot of O_1	
True origin	q.v. county concerned	
False co-ordinates	0, 0	
Scale factor	1	
Sheet dimensions	six-inch full sheet	$31,680 \times 21,120$ feet
	six-inch quarter sheet	$15,840 \times 10,560$ feet
	1:2500 sheet	$7,920 \times 5,280$ feet

O_1 is the Ordnance Survey standard ten-foot bar
1 foot of O_1 = 0.304 800 749 1 international metre

Co-ordinates of neat lines of significant sheets

All data, with the sole exception of Lundy, are for full six-inch sheets. Data either relate to 'home sheets', containing or embordering in-county origins, or are for relevant sides of sheets at cardinal extremities of counties from which other sheet line co-ordinates can be obtained by addition. Sheet data are tabulated by county names current at the time of early editions.

for	Angus	*see*	Forfar	*for*	Kinross	*see* Fife
	Brecon		Brecknock		Midlothian	Edinburgh
	Bute		Argyll		Moray	Elgin
	Caernarvon		Carnarvon		Salop	Shropshire
	Clackmannan		Perth		Shetland	Orkney
	Cromarty		Ross		Southampton	Hampshire
	Dunbarton		Dumbarton		West Lothian	Linlithgow
	East Lothian		Haddington		Zetland	Orkney

Categories:

A	Superseded by combined county origins		E	Included in final 38
B	Superseded in Replotted Counties operation		I	In-county origin
C	Superseded by mystery transfer		O	Out-county origin (or out-area)
D	Superseded in Close's programme		U	Unified counties origin

map	county	origin	cat.	sheet	west n/l	east n/l	north n/l	south n/l
	ENGLAND (including Monmouth)							
40	Bedford [a]	St Paul's	EO	19	-14,256	+17,424		
				34				+98,208
43	Berkshire	Dunnose	EO	43		0		
				44	0			
				50				+244,240
39	Buckingham	Dunnose	EO	31	0			
				58				+286,480
29	Cambridge	Danbury Ch. Sp.	EO	50	-18,000			
				62				+96,080
12	Cheshire [b]	Nantwich Ch. Tr	EI	56	-5,100	+26,580	+17,580	-3,540
48	Cornwall	Hensbarrow	EI	41	-20,880	+10,800	+14,420	-6,700
	Isles of Scilly			83		-306,000	-133,420	
2	Cumberland	High Pike	EI	47	-28,710	+2,970	+10,230	-10,890
17	Derby	Dunnose	EO	19		0		
				20	0			
				62				+751,120
49	Devon	Rippon Tor	EI	108	-18,180	+13,500	+2,820	-18,300
	Lundy			4A NW		-201,320		+224,580
52	Dorset	Blackdown	EI	46	-26,180	+5,500	+7,520	-13,600
4	Durham	Brandon Down	EI	26	-14,520	+17,160	+10,560	-10,560
42	Essex							
	– *Old Series*	Danbury Ch. Sp.	DI	53	-18,000	+13,680	+11,600	-9,520
	– *New Series*	St Paul's	EO	77	-14,256	+17,424		
				86			+13,728	-7,392

map	county	origin	cat.	sheet	west n/l	east n/l	north n/l	south n/l
37	Gloucester	Dunnose	EO	22A		-63,360		
				78				+286,480
53	Hampshire	Dunnose	EI	98		0	+11,920	-9,200
				99	0		+11,920	-9,200
24	Hereford	Llangeinor	EO	23	+71,240			
				53				+64,560
41	Hertford	St Paul's	EO	46	-14,256	+17,424		
				47				+13,728
28	Huntingdon[a]	Danbury Ch. Sp.	EO	19		-113,040		
				27				+159,440
47	Kent	Hollingbourne	EI	43	-20,280	+11,400	+6,820	-14,300
6	Lancashire[b]	Bleasdale	EI	40	-26,400	+5,280		0
				45	-26,400	+5,280	0	
20	Leicester	Dunnose	EO	52		0		+645,520
				53	0			+645,520
19	Lincoln	Dunnose	EO	24	+31,680			
				151				+730,000
45	London	St Paul's	EI	c	-14,256	+17,424	+13,728	-7,392
44	Middlesex[d]	St Paul's	EO	12	-14,256	+17,424		
				16			+13,728	-7,392
36	Monmouth	Llangeinor	EO	10	+39,560			
				27			+1,200	-19,920
30	Norfolk	Danbury Ch. Sp.	EO	92	-18,000	+13,680		
				110				+222,800
27	Northampton	Dunnose	EO	66		0		+476,560
				67	0			+476,560
1	Northumberland							
	– Old Series	Simonside	DI	44	-12,680	+19,000		0
				53	-12,680	+19,000	0	
	– New Series	Brandon Down	EO	94	-14,520	+17,160		
				114			+10,560	-10,560
18	Nottingham	Dunnose	EO	52		0		+772,240
				53	0			+772,240
38	Oxford	Dunnose	EO	45		0		
				46	0			
				58				+286,480
21	Rutland	Dunnose	EO	12	+63,360			
				15				+687,760
15	Shropshire	Llangeinor	EO	68	+71,240			
				82				+233,520
50	Somerset	Blackdown	EO	94	-26,180	+5,500		
				95				+28,640

map	county	origin	cat.	sheet	west n/l	east n/l	north n/l	south n/l
16	Stafford, *partial*	Stafford Castle	AI	16		0		+60,560
				35	0		+18,320	-2,800
	– full series	Dunnose	EO	54		-63,360		
				75				+645,520
31	Suffolk, *partial*	Otley Church Tr	AI	75	0		-36,640	
				80		0		
	– full series	Danbury Ch. Sp.	EO	71	-18,000	+13,680		
				86				+74,960
46	Surrey	Leith Hill Tower	EI	33	-12,000	+19,680	+21,103	-17
54	Sussex	Ditchling	EI	53	-14,000	+17,680	+7,200	-13,920
26	Warwick	Dunnose	EO	28		0		
				29	0			
				59				+476,560
3	Westmorland	Forest Hill	EI	27	-30,880	+800	+13,120	-8,000
51	Wiltshire	Dunnose	EO	78		-63,360		+117,520
25	Worcester	Dunnose	EO	44		-95,040		
				60				+476,560
7	York	York Minster	EI	174	-14,200	+17,480	+8,600	-12,520
	WALES							
8	Anglesey	Llangeinor	EO	15		-87,160		
				25				+529,200
34	Brecknock	Llangeinor	EO	49	-23,800	+7,880		+22,320
22	Cardigan	Llangeinor	EO	13	-23,800			
				46				+127,920
33	Carmarthen	Llangeinor	EO	19	-23,800			
				58				+1,200
9	Carnarvon	Llangeinor	EO	30		-23,800		
				48				+402,480
10	Denbigh	Cyrn-y-Brain	EI	27	-18,080	+13,600	+14,720	-6,400
11	Flint [e]	Cyrn-y-Brain	EO	18	-18,080	+13,600	+14,720	-6,400
35	Glamorgan	Llangeinor	EI	26	-23,800	+7,880	+1,200	-19,920
13	Merioneth	Llangeinor	EO	39	-23,800	+7,880		
				48				+318,000
14	Montgomery	Llangeinor	EO	51	-23,800	+7,880		+254,640
32	Pembroke	Highgate	EI	39		0	+17,820	-3,300
				40	0		+17,820	-3,300
	The Smalls			31C		-158,400		+17,820
23	Radnor	Llangeinor	EO	21	-23,800	+7,880		
				38				+127,920
5	**ISLE OF MAN**	South Berule	EI	12	-17,380	+14,300	+10,920	-10,200

map	county	origin	cat.	sheet	west n/l	east n/l	north n/l	south n/l
	SCOTLAND							
66	Aberdeen	The Buck	EI	51	-5,280	+26,400	+5,280	-15,840
68	Argyll & Bute	Cruach-na-Sleagh	EU	148	-2,600	+29,080	+10,600	-10,520
79	Ayr	Brown Carrick	EI	38	-25,080	+6,600	+7,020	-14,100
65	Banff	The Buck	EI	42A	-5,280	+26,400	+5,280	-15,840
85	Berwick	Derrington Great Law	EI	15	-21,680	+10,000	+6,120	-15,000
57	Caithness	Dunnet Head	EI	1	-15,880	+15,800	+7,900	-13,220
71	Dumbarton,[f]							
	– Old Series	Dumbarton Cas.	DI	22	-9,900	+21,780	+10,560	-10,560
	– New Series	Lanark Ch. Sp.	EO	24		-83,040		
				25				+60,360
82	Dumfries	Hart Fell	EI	9	-29,680	+2,000	+3,120	-18,000
77	Edinburgh							
	– Old Series	Edinburgh Castle	BI	2	-16,980	+14,700	+17,620	-3,500
	– New Series	The Buck	EO	1			-459,360	
				9	-5,280	+26,400		
64	Elgin	Findlay Seat	EI	13	-7,920	+23,760		0
				18	-7,920	+23,760	0	
73	Fife & Kinross							
	– Old Series	West Lomond	BU	16	-11,080	+20,600	+16,020	-5,100
	– New Series	The Buck	EO	1	-5,280	+26,400	-290,400	
70	Forfar	Craigowl	EI	43	-29,040	+2,640		0
				49	-29,040	+2,640	0	
78	Haddington							
	– Old Series	Traprean Law	BI	10	-28,680	+3,000	+6,600	-14,520
	– New Series	The Buck	EO	1	-5,280	+26,400	-438,240	
62	Inverness (mainld)	Ben Auler	EI	154	-23,760	+7,920	+11,880	-9,240
59	Hebrides [gh]	Cleisham	EI	10		0	+5,280	-15,840
				11	0		+5,280	-15,840
60	Skye [h]	Ben Auler	EO	48		-213,840		
				74			+11,880	-9,240
67	Kincardine							
	– Old Series	Caerloch	CI	10	-29,040	+2,640	+17,160	-3,960
	– New Series	The Buck	EO	3			-36,960	
				12	+26,400			
81	Kirkcudbright							
	– Old Series	Bengray	BI	44	-3,700	+27,980	+1,400	-19,720
	– New Series	Brown Carrick	EO	1			-14,100	
				15	+6,600			
75	Lanark	Lanark Ch. Sp.	EI	25	-19,680	+12,000	+18,120	-3,000
76	Linlithgow							
	– Old Series	Mount Airy	DI	5	-17,160	+14,520		0
				9	-17,160	+14,520	0	
	– New Series	Lanark Ch. Sp.	EO	13	-19,680	+12,000		+39,240

map	county	origin	cat.	sheet	west n/l	east n/l	north n/l	south n/l
63	Nairn	Cairn Glasher	EI	12		0		0
				13	0			0
	Orkney & Shetland[i]							
55b	Orkney	Foula	EO	73A		-23,680	-250,740	
55a	Shetland	Foula	EO	34	+39,680			
				55			+2,700	-18,420
	Fair Isle	Foula	EO	69	+71,360		-211,300	
	Foula	Foula	EI	54	-12,680	+19,000	+8,400	-12,720
83	Peebles	Dunrich	EI	17	-13,200	+18,480	+15,840	-5,280
69	Perth & Clackmannan	Ben Cleugh	EU	133	-31,580	+100	+100	-21,020
74	Renfrew	Broadfield	EI	11	-21,120	+10,560	+19,800	-1,320
	Ross & Cromarty[j]							
61	(mainland)	Scournalapich	EU	113	-23,760	+7,920		0
				120	-23,760	+7,920	0	
58	Lewis[h]	Cleisham	EI	44	0		+5,280	-15,840
	Flannan Is[k]			k				
	Rona[l]			l				
	Sula Sgeir[l]			l				
86	Roxburgh							
	– Old Series	Rubers Law	DI	26	-14,850	+16,830	+2,640	-18,480
	– New Series	Lanark Ch. Sp.	EO	1			+18,120	-3,000
				28A	+107,040			
84	Selkirk	Sandhope Heights	EI	14	-24,420	+7,260	+2,640	-18,480
72	Stirling,[f]							
	– Old Series	Corkmulaw	DI	28	-7,920	+23,760	+5,940	-15,180
	– New Series	Lanark Ch. Sp.	EO	37	-19,680	+12,000		+60,360
56	Sutherland	Ben Clibrig	EI	63	-29,920	+1,760	+3,520	-17,600
80	Wigtown							
	– Old Series	Knock of Luce	BI	18	-8,400	+23,280	+19,420	-1,700
	– New Series	Brown Carrick	EO	2	-25,080	+6,600	-119,700	

[a] Huntingdon (det.) is included with Bedford
[b] parts of Cheshire at 1:2500 are included with Lancashire
[c] St Paul's home sheet was variously, according to series, Middlesex 17, or London 7, K, or 5
[d] administrative county area; for complete ancient county see note (c) and London
[e] including detached parts
[f] Dumbarton (det.): first edition was included with main county, but for subsequent editions it was unified with Stirling (Old and New Series)
[g] here signifies the Outer Hebridean chain from Harris to Berneray
[h] numbered in separate series
[i] divided in 1889 into the separate counties of Orkney and Zetland; thereafter technically unified counties
[j] before 1889 were unified counties, thereafter a single county
[k] inset on Lewis sheet 48, with geographical sheet lines
[l] inset on Lewis sheet 1, with geographical sheet lines

Unpublished

Ordnance Survey 1:2500
Staffordshire and Suffolk partial series

1:2500 mapping of Staffordshire and Suffolk commenced in three localised areas, with individual county sheet lines based upon in-county meridians, but after completion of these areas both counties were re-started on combined county meridians as described overleaf. The original partial series comprised two areas around Cannock Chase and Felixstowe, which were published by parishes, and a third of Hanley borough, as it then was. These series were those recorded as '(fragmentary)' by Richard Oliver in Chapter 5 of *Ordnance Survey maps: a concise guide for historians.*[1]

The indexes below show the relevant sheet patterns. Note that sheets which would only include small areas were usually published on adjoining sheets. No six-inch scale sheets were published on these sheet lines. Further technical information will be found overleaf.

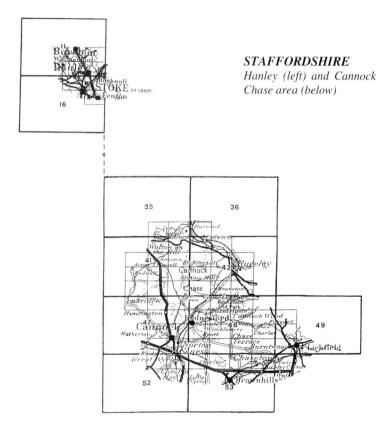

STAFFORDSHIRE
Hanley (left) and Cannock Chase area (below)

---

[1] London: Charles Close Society, 1993. Chapter 7 in the revised and expanded second edition, published in 2005. Reference to these partial series is lacking from Brian's original 1989 articles; his earliest records of them are dated 1993.

SUFFOLK
Felixstowe area
('Felixstow' when
first mapped)

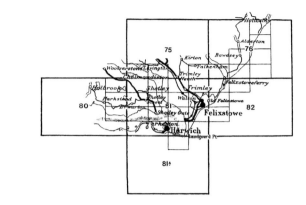

Construction of Ordnance Survey large-scale plans

All Ordnance Survey plans on the scales of 1:2500 and 1:10,560 (six-inch) throughout the British Isles were constructed on Cassini's Projection. This projection is plotted by co-ordinates similar to algebraic x, y co-ordinates, but here measured east or west and north or south from a 'county origin'. Distances are measured true along the central north-south meridian through the origin, and along great circles perpendicular to that central meridian; this procedure creates some distortion in areas away from the central meridian and consequently use of this projection is confined to areas of limited east-west extent.

Hence initially, for the large scales, each county was surveyed and mapped as a separate entity on its own county origin, but after some years the practice was adopted of mapping convenient groups of counties together as 'combined counties', using a single origin and central meridian. Only in Staffordshire and Suffolk was mapping commenced on an in-county origin which was shortly abandoned in favour of a combined counties origin, resulting in the partial series described here. However, there were three phases during which counties which had been completely mapped on their own origins were transferred entirely on to other origins already in use, forming further groups of combined counties.

These various policies brought an overall total of 83 county origins into being, 21 in England & Wales, one in Man, 29 in Scotland, and 32 in Ireland. The origin used for the partial series in Staffordshire was Stafford Castle (Centre), and that for the partial series in Suffolk was Otley Church Tower.

During the nineteenth century many towns were mapped by the Ordnance Survey on various larger scales, and most of these were constructed on the relevant county origins and fitted in to the county sheet line patterns. However, a number of the earliest series were based on individual town surveys and were constructed on local origins, with local sheet-line systems. For completeness it is recorded here that the 1:500 scale plans of Longton published in 1856/57 and of Hanley published in 1865/66 were also constructed on the Stafford Castle (Centre) origin referred to above.

Unpublished, July 2002

Ireland

I derived my series title[1] from Brigadier Winterbotham's 1938 article '150 years and 150 meridians' and so far I have followed him in restricting my coverage to Great Britain and the Isle of Man, but I wish to conclude by highlighting the situation in Ireland, one of those select places fashioned by nature to fit neatly on to the page of an atlas and equally neatly bisected by the meridian of longitude 8° West of Greenwich. This meridian has served as the central meridian for all the national OS projections of Ireland, the Bonne's Projection used for the old one-inch and smaller scale maps, the old Irish Grid on Cassini's Projection used for military versions, and the Transverse Mercator Projection of Ireland and the new Irish Grid which is its visual manifestation. All these, too, have a common centrally-placed true origin, situated just in the waters of Lough Ree, at latitude 53° 30' N, longitude 8° 00' W, forming a distinct contrast to the four national meridians which have been used in Great Britain, Delamere, 4°W, Dunnose, and 2°W.

So, is Ireland then a cartographer's dream? – for small scales maybe, but for larger scales it has been a cartographer's nightmare. Winterbotham refers to 'respectable fully-fledged' meridians; taking these words out of context it could be said that Ireland had only one such, 8°W as above, for all the others were at best provisional. Under pressure to commence large scale surveying in Ireland, the OS started while the primary triangulation was still being completed and well before it was computed and adjusted. Consequently, only provisional values could be determined for the positions of the thirty-two county origins, and no latitude or longitude figures were provided in the sheet borders. But not just their positions were provisional, so were the corresponding directions of true north, and although each county six-inch series could stand on its own as a separate ungraduated block, something had to be done when the time came to reduce the thirty-two blocks into a cohesive whole to draw the one-inch series. So the computing staff at Southampton carried out an intriguing but completely lost 'adjustment' to bring the thirty-two blocks into some sort of sympathy, resulting in another set of provisional geographical positions for the county origins and thirty-two 'county twists' or angles of rotation, each applied to the respective county six-inch block in its entirety. The county twists ranged from 32.82 seconds anti-clockwise to 48.34 seconds clockwise, in the sense to bring the county sheet lines on to true north. No suggestion ever appears to have been made to transfer counties and form combined blocks in Ireland.[2]

So far I have omitted reference to the town plans, and it may have been them and the indeterminate nature of much of their surveying backgrounds that decided Winterbotham to omit reference to Ireland altogether; I am indebted to Professor John Andrews for some additional background to his published works on the subject.[3] It is difficult to decide whether the original manuscript plans would have come within the Brigadier's terms of reference; some fall distinctly short of his requirements whilst the Dublin series was eventually published and would certainly have qualified. Suffice to say here that 125 towns were drawn, probably all on their own meridians, but nineteen of these contained their respective county origins and may therefore be expected to have shared those origins (but I decline to say

[1] The text of this piece, together with the map, originally formed part five of '198 years and 153 meridians, 152 defunct'. The table has not previously been published.

[2] Murphy, Thomas, *The latitudes and longitudes of the six-inch sheet maps of Ireland*, Geophysical Bulletin, 13, Dublin: Institute for Advanced Studies, 1956.

[3] Andrews, J H, *History in the Ordnance map*, Dublin: OSO Phoenix Park, 1974; *A paper landscape*, OUP, 1975.

'confidently expected'). Moving forward to the printed town plans: although they were surveyed independently they were tied in, or more probably forced in, to the surrounding smaller scale work and the majority were on the county origins, but further research is needed to ascertain whether some of the earlier printed plans were oriented on local origins. It therefore appears that the total number of origins or meridians used in Ireland was about 140.

I finish with a Government Health Warning for the unwary – Irish large scale plans can seriously damage your calculations. These plans are drawn to a different module from those in Great Britain, for they are related to the Irish full six-inch sheet size which represents 32,000 feet × 21,000 feet, not immediately obviously different from the 31,680 feet × 21,120 feet of the Great Britain plan sheets.

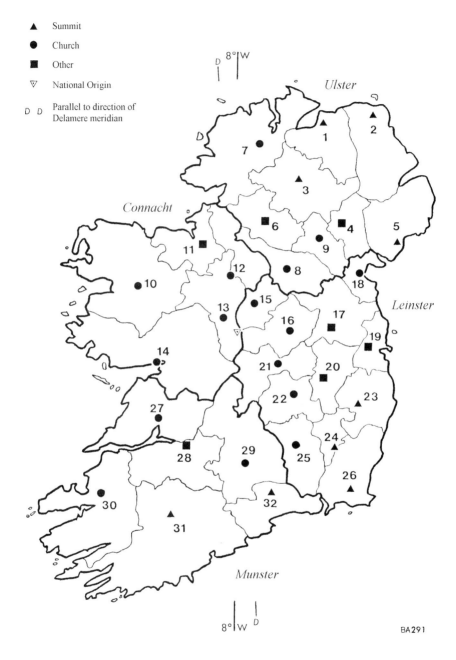

Summit

Church

Other

National Origin

Parallel to direction of
Delamere meridian

8°W

Ulster

1

2

7

3

Connacht

6

4

5

11

9

12

8

10

18

15

Leinster

13

16

17

19

14

21

20

23

22

24

27

25

28

29

26

30

32

31

Munster

8°W

BA291

Ordnance Survey County Series plans – origins and sheet line data

Ireland

Ordnance Survey six-inch scale mapping started in Ireland, before the completion of and well before the adjustment of the Principal Triangulation. Consequently the geographical positions of the county origins, as well as the directions of sheet line north therefrom, were provisional and not consistent. Later, the preparation of the one-inch map required the county plans to be brought into line as far as possible, and an adjustment was carried out at Southampton of which only the results have survived. These are a set of 'county twists' each of which has to be applied to the relevant county series en bloc and has to be utilised in all computations concerning sheet lines.

On completion of the adjustment of the Principal Triangulation geographical positions became available for some of the county origins, in total less than half, leaving the remaining county series rather floating. However, sufficient data is available to enable the present compiler to perform an adjustment correct to about ten feet for the remaining county origins, and the results are included in the listing below, it should be noted that the names in use are those which were current in the mid-nineteenth century.

The defining elements of the Irish county series plans are the same as those for the plans subsequently drawn for Great Britain, although the sheet sizes are different but not obviously different, and some sheets are displaced from the general sheet line blocks.

County twist: when positive, sheet line is *east* of true north at the county origin; when negative is *west*.

Sheet dimensions: six-inch sheet   32,000 × 21,000 feet
  1:2500 sheet   8,000 × 5,250 feet

map county	origin	latitude N		longitude W		county twist	home sheet with its neat line co-ordinates			
		provisional	adj *	provisional	adj *			west	east	north	south
2 Antrim	Knocklayd	55°09' 42".69	43".575	6°15' 00".71	00".168	-11".90	8	-31,400	+600	+15,840	-5,160
							¶ 1	-11,082	+20,918	+55,540	+34,540
4 Armagh	Armagh Observatory Transit	54 21 10.53	10.507	6 38 55.82	55.825	-18.76	12	-17,000	+15,000	+11,000	-10,000
24 Carlow	Mount Leinster	52 37 03.47	04.3	6 46 46.96	46.1	+16.41	23	-14,600	+17,400	+200	-20,800
8 Cavan	Cavan Church Spire	53 59 37.21	37.2	7 21 35.31	35.3	-13.27	20	-24,400	+7,600	+17,700	-3,300
27 Clare	Ennis Church Tower	52 50 44.07	44.7	8 58 51.44	50.8	-10.19	33	-27,200	+4,800	+11,800	-9,200
31 Cork	Mount Hillary	52 06 34.72	35.350	8 50 20.50	19.844	-4.69	31	-28,000	+4,000	+17,700	-3,300
7 Donegal	Letterkenny Church Sp.	54 57 02.84	03.590	7 44 18.53	17.937	-14.48	53	-13,000	+19,000	+15,400	-5,600
5 Down	Slieve Donard	54 10 48.28	48.3	5 55 12.05	12.1	-18.10	49	-18,200	+13,800	+16,700	-4,300
19 Dublin	Dublin (Dunsink) Observatory Transit	53 23 13.00	13.0	6 20 17.51	17.5	-28.80	14	-4,300	+27,700	+17,400	-3,600
							¶ 9	+59,700	+91,700	+48,900	+27,900
							¶ 16	+59,700	+91,700	+9,600	-11,400

No.	County	Location										
6	Fermanagh	Devenish Round Tower	54 22 12.97	13.7	7 39 19.45	18.9	-17.99	22	-4,200	+27,800	+5,820	-15,180
14	Galway	Galway Church Spire	53 16 20.49	21.070	9 03 11.60	10.981	+3.02	94	-11,500	+20,500	+3,400	-17,600
30	Kerry	Tralee Church Spire	52 16 13.61	14.200	9 42 11.99	11.296	-3.11	29	-13,200	+18,800	+15,200	-5,800
20	Kildare	Kildare Round Tower	53 09 27.68	28.493	6 54 41.86	40.871	-17.73	22	-23,600	+8,400	+15,600	-5,400
25	Kilkenny	St Mary's Church Spire, Kilkenny	52 39 04.41	05.1	7 15 07.00	06.4	-17.59	19	-20,400	+11,600	+9,500	-11,500
21	King's Co.†	Tullamore Church	53 16 17.64	18.298	7 28 50.91	50.339	-24.49	17	-9,000	+23,000	+11,500	-9,500
12	Leitrim	Carrick-on-Shannon Church Spire	53 56 48.43	48.4	8 05 37.74	37.7	-26.82	31	-11,000	+21,000	+5,000	-16,000
28	Limerick	Rice's Monument (Limerick)	52 39 27.61	28.2	8 37 40.72	40.1	-4.34	5	-15,500	+16,500	+16,400	-4,600
1	Londonderry	Benevenagh	55 06 49.03	49.862	6 54 56.53	56.054	-15.62	6	-1,700	+30,300	+12,300	-8,700
15	Longford	Longford Church Spire	53 43 51.35	52.0	7 47 57.85	57.3	-14.03	13 ¶	-28,000	+4,000	+6,200	-14,800
								1	+20,000	+52,000	+90,200	+69,200
18	Louth	Dundalk Church Spire	54 00 29.73	29.7	6 24 03.43	03.4	-24.00	7	-12,100	+19,900	+9,100	-11,900
10	Mayo	Castlebar Church Tower	53 51 15.07	15.120	9 17 58.72	58.698	-48.34	78	-22,900	+9,100	+6,200	-14,800
17	Meath	Wellington Testimonial (Trim)	53 33 06.13	06.9	6 47 35.17	34.3	+32.82	36	-22,100	+9,900	+7,900	-13,100
9	Monaghan	Monaghan Church Twr	54 14 53.06	53.1	6 58 07.38	07.4	-26.81	9	-22,000	+10,000	+14,000	-7,000
22	Queen's Co.‡	Maryborough (Portlaoise) New Church	53 02 01.31	02.0	7 18 10.24	09.5	-4.69	13	-8,000	+24,000	+14,000	-7,000
13	Roscommon	Roscommon Church	53 37 43.73	43.7	8 11 24.87	24.9	-30.27	39	-26,800	+5,200	+12,000	-9,000
11	Sligo	Cooper's (Markree) Observatory Transit	54 10 30.17	30.131	8 27 24.93	24.581	-30.48	26	-29,000	+3,000	+3,800	-17,200
29	Tipperary	Cashel Church Spire	52 30 53.48	54.144	7 53 06.71	06.045	-9.47	61	-3,000	+29,000	+10,500	-10,500
3	Tyrone	Mullaghcarn	54 40 26.38	27.2	7 12 31.47	31.0	-19.63	26	-23,900	+8,100	+6,600	-14,400
32	Waterford	Knockanaffrin	52 17 18.46	19.135	7 34 53.71	52.941	-15.85	6	-17,100	+14,900	+5,200	-15,800
16	Westmeath	Mullingar Church	53 31 27.91	28.6	7 20 20.88	20.1	-7.22	19	-11,000	+21,000	+14,700	-6,300
26	Wexford	Forth Mountain	52 18 56.00	56.829	6 33 42.37	41.461	-5.19	42	-6,200	+25,800	+3,200	-17,800
23	Wicklow	Lugnaquilla	52 57 59.94	00.710+	6 27 50.23	49.370	-16.40	28	-23,400	+8,600	+50	-20,950

† now Offaly   ‡ now Laois   + i.e. 52° 58' 00".710   ¶ sheet displaced from the regular sheet pattern for the county

* adjusted: if position of origin is given to three places, adjusted by OS; if to one place, adjusted by Brian Adams, 2004

Indexes to the 1:2500 and six-inch scale maps: Scotland

The user of this book may well be familiar with David Archer's earlier reprint of the equivalent volume for England and Wales, or indeed with the Ordnance Survey's original issue of that work. If not, I hope he or she will bear with me in making some initial reference to those publications.[1] The Ordnance Survey original necessarily depicted the situation at a particular point in time, the date of publication, but the Archer reprint copes with the subsequent re-issue of the plans of two English counties on different sheet lines by the inclusion of additional indexes, whilst the fact that some of the London plans carried up to five different numbers during their lifetimes is catered for by Richard Oliver's feature thereon. But, for example, the user will look in vain for any reference to Kent 85 where formerly plans had depicted an area transferred to Sussex in 1895, nor to Lancashire 111A including an area transferred from Cheshire in 1931. Such changes could be inferred by anyone with an appreciation of the history of administrative areas, but some explanation would be needed for the existence of two first editions of 1:2500 plan Suffolk 82.5, one of them mapping an area twelve miles distant from the one shown in the book. Whilst changes in the patterns of county series plans of England and Wales not catered for in David Archer's reprint only affect a quite small proportion of those plans, such changes were considerably more widespread north of the border, and the present reprint contains a quantity of additional material in order to present as comprehensive a picture as possible of the changing situation during the time of county series plans in Scotland. Before summarising the story of these plans, I describe some of the operations involved.

Cassini's Projection

To understand fully the general arrangement of the Ordnance Survey county series plans, and some of the changes which took place in them, it is necessary to have an appreciation of the basic principles of the map projection upon which they were all constructed, Cassini's Projection. Let me, hopefully, explain the easy part first; the reader has probably seen at the front of his atlas, or elsewhere, some complex diagrams illustrating the principles behind various map projections. But these principles are no use in themselves to the practical cartographer; to draw his flat map he needs those principles translated into x and y measurements so that he can construct the framework of his map, starting probably from the south-west corner, and will then fill in the detail using similar x and y co-ordinates to fix the salient points. The principles of the projection are used to derive the projection formulae, which convert the spatial relationship between any given point and a selected origin into the required x,y co-ordinates. The reader will then see that, if a map series is involved, the sheet lines of the series can be simple x and y co-ordinate lines related to a convenient origin, and this was how the county series plans were set out.

As I describe below, the Ordnance Survey large scales were surveyed and mapped as individual counties or groups of counties, and a local county origin was selected at a triangulation station near the north-south centre line of the county or group; the meridian of longitude through the county origin became the central meridian of the projection for that particular county/group. Since confusion has been caused by different geodesists choosing different directions for x and y, I always follow the unequivocal practice of using e and n;

---

[1] I apologise for perforce using the simple 'he' in the remainder of this introduction.

however, in contrast with the familiar National Grid where use of a false origin renders all eastings and northings positive, in the county series both co-ordinates could be either positive (east or north) or negative (west or south), and were in feet. Neither Cassini grid nor Cassini co-ordinates were printed on the map, but in modern terms the sheet lines of all the county series plans were un-numbered grid lines of a county Cassini grid.

I now come to the specific principles of Cassini's Projection, first noting that for many years the Ordnance Survey did not use a proper name for this, their predominant projection for well over a century, but always referred to it as Projection by Rectangular Spheroidal Co-ordinates. Cassini's is technically a transverse cylindrical projection, which means that instead of being founded upon the Equator, as are many small scale projections, it is founded upon a meridian of longitude. This meridian is chosen to pass centrally down the area which is being mapped, and an origin of co-ordinates is positioned arbitrarily upon it to provide a convenient zero for measurements. Distances from the origin along the central meridian are plotted true to scale along the central meridian of the map; distances along great circles cutting the central meridian at right angles are also plotted true to scale, along straight lines perpendicular to the central meridian on the map. It may help to visualise the situation to see that those great circles bear the same relationship to the central meridian as do meridians of longitude to the Equator. The result of this method of projection is that distances parallel to the central meridian are increasingly too great on the map the farther they are away from the meridian, whilst the north-south scale becomes increasingly larger than the east-west scale at the same point. It is the element of distortion so introduced on to the map that restricts the use of this projection to areas of relatively narrow extent in longitude, of three or four counties width in the case of the large scale plans, although a single projection was used for the whole of England and Wales on the one-inch scale and subsequently extended to cover Scotland. It should be noted that this distortion in the projection is significantly more apparent and vexatious to the surveyor than to the cartographer, whilst the map user who may be totally unaware of the effects of the projection is apt to make unwarranted assumptions about directions and distances.

Transfer of counties

This operation, which occurred in several Scottish counties, as I detail further on, consists of transferring the survey, co-ordinates and large scale mapping of an entire county on to a different county origin. It will be apparent from my description of the projection above that this involves moving the line of zero distortion from one central meridian to another, thereby decreasing the scale on one and increasing the scale on the other, as well as converting the old central meridian from a straight line to a slightly curved one. Thus the transfer is not a simple arithmetic operation, and is effected by converting the latitudes and longitudes of the primary and secondary triangulation stations into e,n co-ordinates related to the new origin, and tying in the detail survey to their new plotted positions. This also produces a new set of sheet lines oriented at a small angle to the original set; this angle, known as the convergence of the meridians, in fact varies very slightly across the area of the county.

As the new origin for the transferred county would already be in use as the origin for another county, and the existing sheet line system for the latter county was extended to cover the transferred county, there was inevitably an overall displacement of the sheet lines as well as the small rotation mentioned. It was a regular practice of the Ordnance Survey, occurring at some time at almost every scale, that when a map series, for whatever reason, was recast on a new set of sheet lines, the sheets issued on amended lines were designated 'New Series'; thereafter the superseded sheets would be referred to as 'Old Series' although this term had not, of course, been used on them or about them during their lifetime. So it was with the plans of transferred counties, but the annotation '(New Series)' was usually deleted from the plates when the plans reached a further edition subsequent to the one at which the transfer was effected. In addition, in the later counties transferred under Close's programme (see below) the sheet numbers of the New Series plans were prefixed by 'N'.

The Replotted Counties

This was the title of an unpublished and overwhelmingly condemnatory paper of 1934 reporting on an operation which was duly referred to in the official history as 'one of the Ordnance Survey's worst errors'. This included six Scottish counties plus the Island of Lewis as well as two English counties, all of which had been surveyed and mapped at 1:10,560 (six inches to one mile) before the decision had been taken to increase the basic scale for most of the country to 1:2500 (25.344 inches to one mile or the 'twenty-five inch scale'). When the 1:2500 survey of the remainder of the country was nearing completion the Ordnance Survey was struck by one of its periodic attacks of economy, and an extraordinary decision was taken in 1886 largely to construct 1:2500 plans for the original 'six-inch counties' from the old 1:10,560 surveys, with minimal re-survey at 1:2500 where absolutely necessary. Every right-thinking mind will boggle at this. By the time this operation reached the Scottish counties concerned in 1892, the numerous representations against it from those involved in the actual process had been ignored, and it was completed fairly rapidly. Looking at those replotted counties in Scotland, rather more than one quarter of their total area was uncultivated ground which was not mapped at 1:2500 scale; of the areas mapped at 1:2500, less than one per cent was reduced to that scale from the very large scale town plans and some twelve per cent was newly surveyed at 1:2500, leaving an incredible 87 per cent of these areas which was drawn at a scale more than four times that at which it had been surveyed. The mind continues to boggle even more! As I record below, the opportunity of the replotting was taken to transfer all these counties, excepting the Island of Lewis, on to the existing origins of other counties.

Boundary changes

The patterns of large scale county plans were also affected by operations of a completely different nature to those described above; I refer to alterations in the county boundaries, of which a high proportion thankfully took place over a short period of time. Some of the old Scottish counties had had some distinctly convoluted boundaries and detached portions, the extreme example being that of Cromartyshire which consisted of twenty-three separate parts scattered through the northern half of mainland Ross-shire (see page 95). The latter situation was resolved by the Local Government (Scotland) Act 1889 which created the new county of Ross and Cromarty, at the same time separating Orkney and Zetland into distinct counties; the Act also established the Boundary Commissioners for Scotland, one of whose duties was to review the county boundaries generally. By 1892 over fifty recommendations for the tidying up of boundaries had been made and put into effect, additional to the absorption of the twenty-three parts of Cromarty. These were to result in the addition or subtraction of a number of the large scale plans to or from their respective county patterns, with in some cases the renumbering of certain sheets.

County Series plans in Scotland

Why county series plans? The idea may seem a bit odd today, with the vast array of Ordnance Survey maps across the whole of Great Britain all fitted neatly together with National Grid sheet lines. But when the Survey first commenced operations at the scale of six inches to one mile, such an array would have been beyond anyone's comprehension. Larger scale mapping had hitherto been on an estate, parish or county basis, usually with little thought of completing the work to the surrounding rectangle (even if one was present). And when the first six-inch maps were drawn, in Ireland, there was no completely observed national triangulation, let alone a computed one, to provide a framework to which a national series of plans could be fitted. So these first plans followed the familiar practice of being prepared as parish maps, and it was only after half a county had been so published that the six-inch was promoted to a county map, each county being surveyed as an entity with its map divided into a rectangular pattern of sheets, completed to their limits within the county but with areas outside the county left blank (with the exception of certain cross-boundary units). Nevertheless the parish map was reverted to when the production of 1:2500 plans was commenced in Great Britain, even though their sheet lines conformed to the county systems. I add for clarification that in the early days of the one-inch map of England and Wales (once thought of as eventually extending into Scotland), although it was being prepared as a national series, it was only indirectly connected to the national triangulation, and the latter had only been adjusted on a disjointed plane basis. Consequently the southern four-fifths of the Old Series maps were not constructed on a single rectangular sheet system.

So it was, that when six-inch mapping crossed the water to Lancashire and Yorkshire, and thence across the border into Scotland, it was carried out on a county basis on Cassini's Projection as described above, and thus it remained together with subsequent larger scale mapping at 1:2500 until the advent of the retriangulation and the National Grid, both related to the Transverse Mercator Projection, in the periods either side of the Second World War. There were, however, departures from the principle of every county being surveyed and mapped as a single entity; the first areas mapped at the six-inch scale in Scotland, commencing from 1843, were the counties of Edinburgh, Haddington, Kirkcudbright and Wigtown, and the Island of Lewis, each of which was individually treated on its own county origin, plus the counties of Fife and Kinross. In a type of operation only found in Scotland these two were treated as what I term 'unified counties'; the survey and mapping of the pair of adjoining counties were referred to a single county origin and the maps were laid out as a single continuously-numbered series, and depicted on a single sheet index. In 1854 surveying began at 1:2500 in Scotland, initially in the counties of Ayr and Dumfries, and (with some interruption during the period known as the 'Battle of the Scales') was thereafter extended to all built-up and cultivated areas of the country. Each twentyfive-inch plan covered one-sixteenth of the area of a six-inch sheet, the sides of the latter being divided into four each way, but as remarked above, the 1:2500 mapping was originally published by individual parishes and continued thus until about 1872. During this second phase of county series mapping the pairs of counties of Argyll and Bute, Perth and Clackmannan, Ross (mainland only) and Cromarty (then still separate) were all surveyed and mapped as unified counties; also during this period Orkney and Zetland were treated as a single unit, not then having been separated by the 1889 act mentioned above, and so after the implementation of that act they effectively appeared as unified counties.

With the latter exception the unified county pairs had included one relatively small county, which it was an evident convenience to survey in concert with a larger neighbour. But it was also becoming clear that surveying counties on a totally individual basis was producing problems along the county boundaries, where the adjoining surveys simply did not fit together satisfactorily, an effect which was magnified by the fact that the methods of adjustment of the surveys tended to push the inevitable errors out to the edges of the areas. So it was that 'combined counties' came into being, as distinct from 'unified counties'. Whilst the surveys of combined counties were continuous across the common county boundaries, and they and the mapping were referred to a single county origin with a single overall pattern of sheet lines, each county had its own sequence of sheet numbers. As the cross-boundary sheets within the combined county system were completed to their rectangles with mapping of both or all counties concerned, such sheets were dual- or even treble-numbered with separate numbers in each county series involved. It is apposite to remark here that these multi-numbered sheets cause much trouble to map librarians and their customers in filing, demanding, retrieving and returning to shelves, and all these operations should be approached with circumspection.

The first combined counties to appear in Scotland were Aberdeen and Banff, possibly as much because the shapes of these counties suggested the same triangulation station as the origin for each county as for any other reason; indeed, the fact that the two counties' plans were shown on a single index sheet suggests that they may originally have been intended to be published as unified counties, but were subsequently individually numbered as ordinary combined counties, within two years of the first adoption of the latter arrangement in England in 1863.

I now move to describe the large scale mapping of the Hebridean islands, Inner and Outer, starting with the county of Inverness. With the mainland area of this county nearing completion, surveying commenced on the Island of Skye, but these two areas were effectively treated as distinct but combined counties in that each was given its own separate sequence of sheet numbers. The county's island chain extending from Harris to Berneray was separately mapped and was treated as an individual county combined with the Island of Lewis on that area's early origin, whilst the mainland portion of Ross-shire had been mapped, as previously stated, as unified with Cromartyshire. Lewis was also involved in the next stage of county series mapping, as follows.

I come now to the infamous episode of the replotted counties explained in a previous section. Following the completion of 1:2500 and six-inch mapping of all the other counties in 1890, the 1:2500 mapping of the six counties and Island of Lewis originally published on the six-inch scale only was prepared by the method of replotting plus minimal re-surveying described above. The occasion of replotting the mainland counties was utilised at the same time to transfer these areas on to other existing origins to form two new groups of combined counties. The counties of Kirkcudbright and Wigtown were transferred on to the Ayr county origin, whilst Edinburgh, Fife & Kinross, and Haddington were all transferred on to that currently in use for Aberdeen and Banff. It will immediately be seen that, although the second group overlapped each other across the Forth, there was a gap of over thirty miles between the north of Fife and the south of Aberdeen. But the whole group constructed on the Aberdeenshire origin were truly combined counties; the north-south sheet lines from Banffshire to Edinburghshire were common co-ordinate lines of Cassini's Projection on that origin, whilst the gap between the southern sheet line of Aberdeen 113 and the northern of

Fife 1 was an exact multiple 21,120 feet (or four miles) as it should be. The publication of the 1:2500 plans of the replotted counties took place from 1892 to 1896.

By that time the production of the second editions of the plans of the other counties had commenced, when the amended sheet line patterns referred to in the section on boundary changes were put into effect. There also took place an operation which apparently remained un-noticed in print for ninety years, the transfer of Kincardineshire also on to the much used Aberdeen county origin. In contrast with the replotted counties already mentioned, this was the first transfer of a county fully mapped at both the 1:2500 and 1:10,560 scales. It seems fairly probable that the previous transfer of the Forthside counties was carried out with the transfer of the counties of Forfar and Kincardine also in mind, but nothing about this is known to have survived in print and the transfer of Forfarshire was never carried out. A much smaller transfer, but one which should not be left un-noticed, was that of the detached portion of Dumbartonshire containing the area of Cumbernauld. This had been published together with the principal part of Dumbarton on the first editions of that county's plans, but when the first revision / second editions came along it was published on the relevant plans of the county of Stirling. Thus the detached part was actually transferred on to the Stirling county origin and became unified with Stirling.

The inconvenience of having discontinuity in mapping across many county boundaries continued to pose problems, especially where built-up and industrialised areas crossed those lines and, soon after his installation as Director-General of the Ordnance Survey, Colonel Charles Close decided to institute a rolling programme to transfer one-by-one as many counties as feasible on to other existing county origins. It may however be seen that combining surveys which did not meet together properly involved some degree of fudging (to avoid any of the usual euphemisms), and thus only presented the inherent problems in a new form. Work on Close's programme commenced in 1913, was interrupted by the First World War and finally abandoned, theoretically temporarily, in the most vicious of all bouts of economy in 1919. By that time four Scottish counties had been transferred, whilst the computations had been completed for another two and the preliminary planning of sheet lines had been done for three more. All these transfers were, or were to be, on to the Lanark county origin, which would ultimately have served the whole of southern Scotland up to the line of the Firth and River of Forth and within Loch Long, the entire area comprising a single group of combined counties. The counties fully transferred were Dumbarton, Linlithgow, Roxburgh and Stirling, and those computed were Dumfries and Selkirk, with preliminary work done on Berwick, Edinburgh and Peebles. In the original scheme Ayr, Haddington, Kirkcudbright, Renfrew and Wigtown would have followed on to the Lanark origin, but Close never set out his ideas for northern and western Scotland. It will be noted that Dumbarton and Stirling were now combined counties, along with others, but the detached Cumbernauld area continued to be unified with the Stirling sheet numbering scheme and shown on the Stirlingshire sheet index, having actually been transferred twice. It will also be seen that in the full scheme the four counties in the area already replotted and transferred would have been transferred again.

There only remain two more episodes in the story of the county series large scale plans, the issue of six-inch sheets as 'Provisional Editions' with the National Grid imposed upon them, although it had not been adjusted into sympathy with their graduations, and their final withdrawal and replacement by sheets constructed on National Grid sheet lines. But those who were familiar with them for so long are still very happy to see them, if not in their own collections then in libraries, and this book should be invaluable in helping to seek out particular sheets required. Hopefully too, this introduction may help to explain some of their complex background to the uninitiated.

Brian Adams
Parsons Green
London SW

Further reading

Brian Adams, '198 years and 153 meridians, 152 defunct', *Sheetlines 25* (August 1989), *26* (December 1989), *27* (April 1990). [Some additional references will be found in this series.]

Brian Adams, 'The scattered county of Cromarty', *Sheetlines 29* (January 1991).

D H Maling, *Coordinate systems and map projections*, London: George Philip & Son, 1973. Second edition Oxford: Pergamon Press, 1991.

Richard Oliver, 'The Ordnance Survey and its indexes to large-scale plans', introduction to *Ordnance Survey of Great Britain: England and Wales, indexes to the 1/2500 and 6-inch scale maps*, reprinted Kerry: David Archer, 1991. Reprinted with additional material by Roger Hellyer, 2002.

Richard Oliver, *Ordnance Survey maps: a concise guide for historians*, London: Charles Close Society, 1993. Second edition, revised and expanded, 2005.

Summary survey and revision dates for OS large scales mapping

This table has been compiled from the various official Ordnance Survey catalogues, from official OS Revision Progress diagrams now deposited with the Charles Close Society, from OS annual reports and from the maps themselves. These dates are not always in agreement, and for this reason the information given in the table below may differ in minor particulars from various predecessors. It should be noted that the dates for post-1945 mapping do not necessarily apply to those built-up parts of counties which were mapped at 1:1250 scale some years in advance of the 1:2500 mapping of the county being started. Counties which were incompletely revised are denoted by the suffix 'I'.

Unless noted otherwise below, all surveys or revisions were at 1:2500 scale for cultivated areas and at six-inch (1:10,560) scale for mountain areas. The suffix 'S' has been added where both the cultivated and mountain parts of a county were mapped only at the six-inch scale.

All first edition six-inch maps were published as full sheets, with a map face 36 inches by 24 inches. Most revised six-inch maps were published in quarter-sheets, with a map face 18 by 12 inches, but the full sheets were retained in a few counties; these are indicated by the suffix 'F'. In addition, a few full sheets were published between about 1920 and 1924 in counties mostly published in quarter-sheet form; these have not been indicated here.

Where a county origin was changed, the suffix 'M' has been added to the revision at which the change took effect.

Fuller details of six-inch and 1:2500 mapping dates, and of areas mapped at scales larger than 1:2500, will be found in Richard Oliver, *Ordnance Survey maps: a concise guide for historians*, London: Charles Close Society, 1993; second edition, revised and expanded, 2005.

	County Series survey	First revision	Second revision	Third revision	National Grid survey
Aberdeenshire	1864-71	1899-1901	1923-5 I		1963-82
Argyllshire	1862-77	1897-8	1914-15 IF		1960-82
Ayrshire	1854-9	1893-6	1907-9	1937-8 I	1958-82
Banffshire	1865-70	1900-01	1928-9 I		1963-82
Berwickshire	1855-7	1896-8	1905-6		1962-82
Buteshire	1855-64	1895-6	1914-15 I		1964-80
Caithness-shire	1870-2	1904-5 F			1964-76
Clackmannanshire	1859-63	1895-9	1920 I	1931 I	1960-78
Dumfriesshire	1854-8	1898-9	1916-17 1929-30 I		1958-82
Dunbartonshire					
Main part	1858-61	1894-8	1914 M	1936-8 I	1964-78
Detached portion	1858-61	1894-8 M	1914 M	1936-8	1958-70
Edinburghshire	1850-2 S	1892-4 M	1905-6 1912-13 I	1931 I	1950-82
Elginshire	1866-71	1902-4			1964-81
Fifeshire	1852-5	1893-5 M	1912-13	1924-5 I 1938-43 I	1959-79
Forfarshire	1857-62	1898-1902	1920-3 I		1965-79
Haddingtonshire	1852-4 S	1892-3 M	1906	1932/38 I	1958-82
Inverness-shire					
Mainland	1866-76	1899-1903 F	1929/38 I		1966-80
Island of Skye	1874-7	1898-1901 F			1964-76
Hebrides	1876	1901 F			1968-75
Kincardineshire	1863-5	1899-1902 M	1922-3 I		1963-82
Kinross-shire	1853-4 S	1893-5	1913		1961-79
Kirkcudbrightshire	1845-50 S	1893-4 M	1907-8	1931 I	1958-82
Lanarkshire	1856-9	1892-7	1908-11		1956-82
Linlithgowshire	1854-6	1894-6	1913 M		1953-65
Nairnshire	1866-9	1903-4			1965-79
Orkney	1877-8	1900 F			1965-78
Peeblesshire	1855-8	1897-8	1906		1964-82
Perthshire	1859-64	1894-1900	1930-1 I		1964-82
Renfrewshire	1856-8	1892-6	1908-12	1937-40 I	1959-79
Ross and Cromarty					
Mainland	1868-75	1901-5 F			1964-82
Island of Lewis	1848-52 S	1895-6 F			1966-76
Roxburghshire	1856-9	1896-8	1916-19 FM		1962-82
Selkirkshire	1856-9	1897	1930 I		1963-81
Stirlingshire	1858-63	1895-6	1913-14 M	1938-43 I	1954-79
Sutherland	1868-73	1903-5 F			1959-76
Wigtonshire	1843-7 S	1892-5 M	1906-7		1965-81
Zetland	1877-8	1900 F	1928 I		1966-73

Alterations to sheet patterns

Firstly, to elaborate on the information given on page 1 of the original volume, the publication of six-inch maps in quarter sheets only dated from their printing by lithography from 1881 onwards. Previously they had been published in the full sheet form indicated by the primary numbering system, and such publication was also reverted to for a short period in the early 1920s.

In addition to the complete recasting of whole counties on different sheet lines resulting from the transfer on to new origins,[2] alterations to existing sheet patterns also took place from time to time. It has not been possible within the scope of this book to undertake the extensive research necessary to provide a complete listing of all such alterations, particularly those in 1:2500 plans, but the following information given in terms of full six-inch sheet numbers should provide a useful guide.

Dumbartonshire: the sheet numbers shown on page 53[3] are those of the second edition. As mentioned in the introduction, the first edition sheets also covered the detached part of the county, subsequently unified with Stirlingshire; page 84 shows the first edition sheets covering this part of the county, but it should also be noted that in the first edition the southernmost pair of sheets of the main county area were numbered 28, 29 instead of 25, 26 as shown on page 53.

Changes in sheet patterns due to county boundary changes: note that, in addition to the boundary changes in the 1890s referred to in the introduction, changes to the limits of the city of Glasgow also affected sheet patterns subsequent to the publication of the original volume.

Aberdeenshire	in the first edition there was no sheet 33A, but there was a sheet 41 west of 42.
Argyllshire	additional first edition sheets 1, 2 existed north of 5, 6, with 3 east of 2, and 7, 12 east of 6, 11.
Banffshire	there was an additional first edition sheet 13A west of 13, but no sheet 42A.
Inverness-shire	there was an additional first edition sheet 6 east of 5.
Lanarkshire	there was an additional first edition sheet 54 east of 53, whilst the sheet numbered 54 on page 61 was originally numbered 55. In the Glasgow area new sheets were later added numbered 1A west of 1, and 10A west of 10.
Peeblesshire	there were no first edition sheets 14A or 18A, but there was a sheet 25 east of 24, and the sheets numbered 25, 26 on page 63 were originally numbered 26, 27.
Perthshire	there was no first edition sheet 120A, but there were additional sheets as follows: 136, 137 south of 130, 131; 141 east of 140; 142 south of 140 and 143 east of 142.
Renfrewshire	with changes in the city of Glasgow, sheet 9 disappeared.
Roxburghshire	there was no first edition sheet 24A.

---

[2] In the Ordnance Survey's own usage new county origins were generally referred to as new initial meridians (or new meridians).

[3] All page references in this section are to the original work in which this appeared, and not to the present volume.

Selkirkshire	there were no first edition sheets 9A, 13A, 16, 17A, but there was an additional sheet 22 east of 21.
Stirlingshire	there were no first edition sheets 10A, 11A.

Other changes in the sheet patterns were more apparent than real; for example in Selkirkshire, in addition to the sheets mentioned above there was no first edition sheet 20A, the small area concerned being mapped as an inset on sheet 20. This area was mapped by extrusions on sheets 20NW and 20SW in the edition depicted on page 69, the sheets actually being numbered 20aNE & 20NW and 20aSE & 20SW, and there were over twenty similar instances spread over Scotland. Whilst the user needs to be warned of such cases, there appears to be no point in attempting to detail them.

> *This introduction to* Indexes to the 1:2500 and six-inch scale maps: Scotland, *(Kerry: David Archer, 1993, ISBN 0951757938) is reprinted by kind permission of the publisher.*

Some problems concerning the County Series numbering of the Small Isles came to light in 1994. Brian investigated and placed his findings at the disposal of map librarians in the RGS and the copyright libraries; they are published here for the first time:

At the time of the first large-scale survey of the Small Isles the group was divided between Argyll and Inverness, with only the island of Eigg being included in the latter county. The rest of the group was transferred to Inverness before the first revision was undertaken, but from the start the whole group's maps had been numbered as part of the independent Island of Skye series of sheets. However, the latter situation was not reflected in the first edition sheet headings (other than by the sheet numbers themselves). The strangest thing about the original county indexes was the depiction of sheets 70 and 72 on the two indexes, which in each case omitted the outline of the island which was not in the relevant county.

The Small Isles – Ordnance Survey six-inch first edition
Inverness-shire (Island of Skye) sheet numbers and headings [1]

Island of Canna, Argyllshire	Sheets LIII & LIV
Islands of Canna & Sanday, Argyllshire	Sheet LIX
Islands of Rum, Sanday &c.	Argyllshire sheet LX
Island of Rum	Argyllshire sheet LXI
Island of Òigh-sgeir &c, Argyllshire	Sheet LXIV
Island of Rum	Argyllshire sheet LXVI
Island of Rum	Argyllshire sheet LXVII
Island of Rum	Argyllshire sheet LXIX
Islands of Rum & Eigg, Argyllshire & Inverness-shire	Sheet LXX (acc.1880)
Islands of Rum & Eigg	Argyllshire & Inverness-shire sheet LXX (acc.1884)
Island of Eigg	Inverness-shire sheet LXXI
Islands of Muck & Eigg, Argyllshire & Inverness-shire	Sheet LXXII
Island of Eigg, Inverness-shire	Sheet LXXIII
Island of Muck, Argyllshire	Sheet LXXIV

---

[1]   Headings and other information taken from the holdings of the Royal Geographical Society.

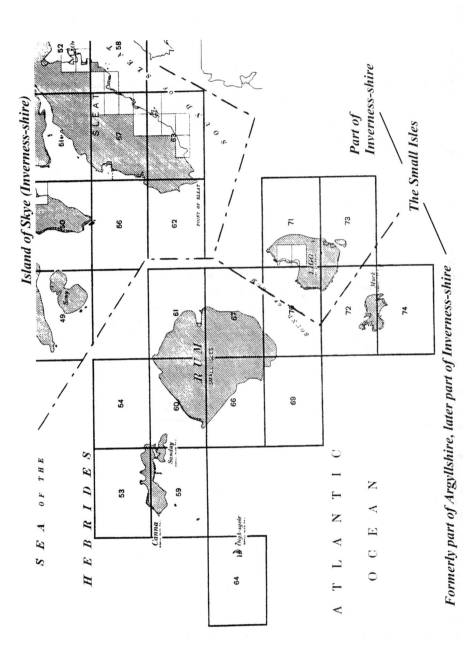

Island of Skye (Inverness-shire)

S E A   O F   T H E

H E B R I D E S

A T L A N T I C

O C E A N

Part of Inverness-shire

The Small Isles

Formerly part of Argyllshire, later part of Inverness-shire

The scattered county of Cromarty

Readers familiar with nineteenth century maps of Scotland will know that the old county of Cromarty consisted of a number of separate portions scattered through the northern half of mainland Ross-shire; they may not have realised that it totalled as many as twenty-three individual areas. In accordance with Ordnance Survey standard practice the principal portion of the county was designated 'CROMARTYSHIRE' and the remainder were 'CROMARTYSHIRE (Detached No.1)' to 'CROMARTYSHIRE (Detached No.21)'; the arithmetic is not in error, please read on. In the peculiar circumstances of this county the principal portion, that containing the town of Cromarty, was not the largest but still the second in area although only one-sixth the size of the largest portion. The latter, Detached No.10, contained the district of Coigach to the north-west of mainland Ross-shire.

By various provisions of the Local Government (Scotland) Act 1889 the counties of Ross and Cromarty were united, those of Orkney and Zetland were separated, and the Boundary Commissioners for Scotland were established inter alia to review county boundaries generally. And so in 1891 W & A K Johnston were able to publish their *Map of Scotland showing the new county boundaries*, which depicted seventy-six areas transferred between counties, although some had not then been confirmed. Comparing the former areas of Cromartyshire shown on this map with the OS six-inch first edition index, I was surprised to come across 'Det. No.14a.' on the latter with no close affinity to 'Det. No.14', and then established that there was no 'Det. No.13' on this index.

I first supposed that this represented a most unusual intrusion of superstition into OS practice, and an equally unusual substitution of 14a instead of 12a for the unlucky 13. However, on examining the individual six-inch sheets I discovered 'CROMARTYSHIRE (Detd. No.13)', with the different abbreviation then current on the actual sheets, to be a small area not labelled on the index sheet. In fact it was an area of only fifty-two acres on the southern tip of Gruinard Island, which undoubtedly experienced more than its share of bad luck when it was selected to be infected with anthrax during the Second World War.

Whilst the intrusion of Det. No.14a into nos 1 to 21 explains the arithmetic, its existence defies reasonable explanation, and it seems to me could only have resulted from major human error, such as duplication, in the original numbering of the detached parts. Does anyone know who numbered these intriguing areas and at which stage in the mapping process? And which authority area had the greatest number of 'Dets.'? - the parish of Glaisdale in the North Riding of Yorkshire formerly ran to twenty-six detached parts, or twenty-seven portions in all; any advance?

Sheetlines 29, January 1991

96

The Landmark Information Group

Very wide-awake members reading Richard Oliver's piece on Landmark in *Sheetlines 45* should have realised that something had been left unsaid. As Richard stated, each Landmark 'SiteCheck' report includes the relevant historic and current Ordnance Survey large-scale mapping of the site and a surrounding buffer zone, in other words a succession of congruent extracts from county series and National Grid plans, usually at 1:2500. But not only are these two classes of plan on different projections, one without and one with a scale factor, more significantly they are constructed on different origins and are therefore on differing orientations. For any one place two different county origins can be involved, whilst the extreme case with three county origins concerned as well as the National Grid happens to be the significant development area of Cumbernauld.

The cartographers of my generation could easily cope with such differently constructed plans by some juggling on a light-table, but at the present time, with the old and new plans digitised and contained in Landmark's computer's memory, the juggling has to be carried out by the computer also. Landmark, led by their Operations Director, former military surveyor James Cadoux-Hudson, were able to devise the software necessary to perform this operation, but they did not have the large quantity of basic data required to put the operation into practice for the fifty-one origins involved. Having found that the Ordnance Survey were unable to supply all this data, James Cadoux-Hudson did the correct thing; he asked David Archer, as publisher of reprints of the county plan indexes, whether he knew all about the county origins.

"No," said David, "but I know a man who does." And in no time at all I was discussing Landmark's requirements with James. He was evidently relieved to learn that I could not only supply all the data for every county origin, including those superseded as long ago as 1875, but that I could also supply the sheet line construction data for all 109 separate county blocks which make up the OS county series plans. I should make it clear that all the above refers to Great Britain and the Isle of Man only,[1] and also that, whilst all the necessary material was contained in my note-books, that relating to the most abstruse superseded series had not been codified and required a short batch of computing before sending it on to Landmark.

So it was that, when I subsequently visited Landmark's headquarters, I was pleased to find that my good name had preceded me and I was welcomed by their staff; and there are two footnotes to this story. Firstly, that we are able to welcome James Cadoux-Hudson into the ranks of the Charles Close Society members; and secondly, that in his address before the Society Annual General Meeting, Professor David Rhind remarked that he was surprised to hear that Brian Adams had been of considerable help to Landmark because he thought that he was a pop-star. But as I pointed out to the Director General, that one can't even spell it right!
(With apologies to any CCS members named Bryan.)

Sheetlines 46, August 1996

---

[1] It was not to remain so; by 2004 Brian had completed the mammoth task of computing comparable data for all 3. separate county origins in Ireland.

The British Isles - how many?

Commander Peter Beazley, formerly Territorial Waters Officer at the Hydrographic Department, was a vital member of the United Kingdom team at the extended series of United Nations Law of the Sea discussions and conferences 1970-1982. In 1974, when the rights and liabilities of archipelagic states were much under discussion, he asked the Charts (Whitehall) unit, headed by the present author, how many islands there were in the UK, and the drawing offices turned temporarily into counting houses. In November 1989 *The Guardian*, eight years after *Sheetlines*, commenced a 'Notes & Queries' feature and the following summer a reader enquired 'How many islands are there in the British Isles?' Building on the previous experience I was able to provide an answer which was published in slightly edited form on 25 June 1990.

As the majority of the counting was carried out on one-inch maps, I offer the results of these exercises to the Charles Close Society with an extended explanation. First it has to be decided what is meant by an island; in Law of the Sea matters the concern is with the 'juridical island', which is basically any detached piece of natural land, however small, which is above the level of mean high water spring tides. Members will recall that this is the level which is depicted as the firm coastline on OS maps of Scotland, whereas on those of England, Wales and Ireland the firm coast follows the line of mean high water of average tides. As the islands are in the true sense of the word innumerable, and the totals will inevitably be rounded off, there is no practical objection nor practicable alternative to doing the count on the one-inch maps of all the countries.

Before leaving the area of laws of the sea it may be of interest to refer to the measurement of territorial seas and similar maritime zones. The subject is somewhat involved, but where there are no complicating factors present such zones are measured outwards from the low-water line, specifically the low-water line as marked on the large-scale charts of the relevant state. Here we have the intrusion of the common sense of the mariner into international conventions, for again there is no practicable alternative, even though the defined level of the charted low-water line varies between places. This level is the same as the chart datum, which should conform to the general principle of being a level below which the tide will but seldom fall.

The main possible types of island are probably better understood from the illustrations than from verbal explanations. For *Guardian* readers I thought that the initial emphasis would best be on those which were islands at all states of the tide, and I then gave a further figure for those islands which were additional at high tide, but it must be appreciated that as the tide sweeps its way around our coasts the latter do not all appear as separate islands at the same time. When they are not islands they are attached to the mainland, or to other islands, by exposed foreshore. I did not attempt to enumerate the third basic type, the low-tide elevations, although many of these would be regarded as islands by the general observer when exposed by the falling tide.

Practicability in counting, as well as a guess at the reader's concept of an island, demanded that a minimum size be adopted; I set this at half an acre or 0.2 hectare. Taking it for granted that islands in lakes or rivers are not to be regarded as separate islands in this context leads to the question of demarcation between rivers and sea. A purely theoretical approach might take all tidal rivers to be arms of the sea, but I doubt if the average person would expect to count in an island many miles from open water. The distinction between

98

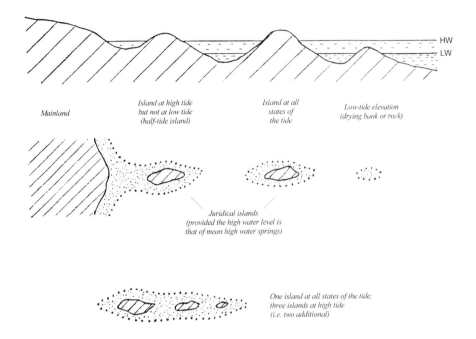

Mainland

Island at high tide
but not at low tide
(half-tide island)

Island at all
states of
the tide

Low-tide elevation
(drying bank or rock)

Juridical islands
(provided the high water level is
that of mean high water springs)

One island at all states of the tide;
three islands at high tide
(i.e. two additional)

riverain and estuarial waters can be discussed at length, but in this operation my own subjective assessment found hardly any instances of doubt. As a matter of interest I have included the figures for islands in tidal rivers in the accompanying table of results. The table gives a general breakdown by countries, but their figures should basically be regarded as steps towards the totals and not as accurate to the degree quoted.

The counting process so far has thus comprised the definition of types and setting of constraints, as well as the adoption of methods to ensure that islands are not counted twice, neither on sheet overlaps nor when they lie across sheet joins. But the number of inhabited islands is also of obvious interest and these present a greater problem; they cannot be specifically identified from any map. All the map shows are those islands which have buildings on them, and in the absence of any other information it is only possible somehow to estimate how many of these are inhabited. Again I made subjective assessments of a, islands which were certainly or almost certainly inhabited, and b, those which might be inhabited, and the figures which appear in the final column of the table were formed by $a + \frac{1}{2}b$. Confirmation of the validity of this empirical method comes from a note in a miscellaneous folder in the Royal Geographical Society Map Room, which says that according to the 1951 census the number of inhabited islands around the coasts of the UK was 187 (the OPCS was unable to find me a more recent figure); if this includes the Crown Possessions as do the overall census returns, it is remarkably close to my figure of 185[q]. I should add that the numbers of inhabited islands in the table are of all islands at high tide, whereas the figures I quoted in The Guardian as inhabited were for islands at all states of the tide only.

Islands of half an acre or more in area

Numbers of islands:-	at all states of the tide	at high tide only	total at high tide	in tidal rivers, etc.	inhabited (estimated)
England	220	500	720	990	31[p]
Wales	120	140	250	90	11[p]
Scotland	3,080	4,180	7,260	200	121[p]
Northern Ireland	40	80	120	20	7[q]
Crown Possessions *	110	160	260	-	14
Republic of Ireland	870	1,020	1,890	210	105[q]
UK & Crown Possessions	3,600	5,100	8,600	1,300	185[q]
British Isles	4,400	6,100	10,500	1,500	291

* *Isle of Man and Bailiwicks of Jersey and Guernsey*
[p] *plus portion of Great Britain*
[q] *plus portion of Ireland*

Note: figures left of rule are rounded and do not necessarily total. Those above rule are rounded to nearest 10; those below to nearest 100.

Finally I refer back to the third column in the table, total islands at high tide; it is these which will be shown as islands on a map or chart, but the actual number marked will depend on the scale of the graphic. My recorded total of 10,500 of half an acre or over on one-inch maps suggests that the overall total of islands marked on this scale, down to mere dots, would be 20,000 or more, and the number on the large-scale plans even higher. But it is salutary to compare these true figures with the numbers shown on smaller scales; thus Admiralty chart 2, the old soundings in fathoms version on a scale of 1:1,686,700, has about 1,720, Bartholomew's *Times Atlas* map on 1:2,500,000 has 428, and Philip's *Encyclopaedia Britannica* map on 1:4,500,000 (Channel Islands on 1:5,000,000) has 202. So, how many islands are there in the British Isles?

'It depends what you mean...?'

Sheetlines 29, January 1991

Brian responded a second time to the question "Exactly how many islands make up the British Isles?" which came up again in the "Answers to correspondents" column in the Daily Mail, 18 July 1995. He was displeased when a crucial line of his answer, in which he detailed the above individual totals for Scotland, Wales, Northern Ireland and the Republic of Ireland, was omitted!

A related piece of research undertaken by Brian was revealed to us by a former member of the UK Hydrographic Office, who reported: "Many years after he retired I discovered Brian was the man who calculated the length of the coastline of the UK, with and without offshore islands. I believe these distances are still quoted by the UKHO but I doubt if anyone remembers from where they originated."

Brian calculated the total length in 1970, using a uniform procedure throughout, as 8,590 miles, including the Channel Islands; he revised his report in 1977, not to adjust the figures, but to take account of 1974-5 boundary changes.

Combined sheets of the Irish one-inch coloured edition

Quite a number of the more senior members of our Society will remember the late Stanley Crowe, a leading antiquarian bookseller specialising in topographical literature and maps, who long occupied premises in a Bloomsbury basement after being bombed out from the City. Many years ago I was fortunate to acquire from him a complete set of the coloured edition one-inch maps of Ireland bound in three volumes; two years later he advertised another complete set at exactly twice the price. However, it was somewhat confusing that the various combined sheets in my set differed in several respects from those depicted on the only series indexes that I could find, and eventually I had to construct my own index to my set.

Having offered to write something about this for the editor's proposed Irish-oriented number, I examined the Royal Geographical Society's holdings of this edition, only to find that my set occupied an intermediate position between the array of coloured sheets as originally published and the final pattern of only a few years later. The differences between the original and later groupings of combined sheets are illustrated by the accompanying map.

The first sheets of the coloured edition were published early in 1902 and the first combined sheet, 201 & 202, appeared in May of that year. It was not until October 1903 that a second combined sheet was published, 71 & 72, followed by 49 & 50 in December and 37 & 38 in March 1904. These four sheets all comprised basic sheets extended eastwards to include relatively small portions of land. Meanwhile the coloured edition had first been completed for the south-eastern quarter of Ireland, followed by the south-western sixth and the north-eastern quarter, and it was only with the completion of the remaining north-western and western parts during 1905 and the first months of 1906 that combined sheets of varying shapes became almost universal in the coastal areas. It is evident from the writings of Professor John Andrews that this surge of interest in combined sheets occurred with the transfer of Irish coloured work from Dublin to Southampton. I should make it clear that I am only concerned here with the coloured sheets of the regular series, and not with special district sheets inspired by the Dorington Committee report.

Subsequently, during 1906 and 1907, quite a number of the early-issued sheets were re-prepared, and during the course of this work many of the original basic sheets around the south-west and south coasts were replaced by additional combined sheets, together with some re-arrangement of the combined sheets on the coast of Ulster, as shown on the map. The numerical designations of the combined sheets followed the rule (with one exception, later corrected) that where portions of a basic sheet fell on more than one combined sheet, then the basic sheet number was qualified by 'Part of' or 'Pt. of' in the combined sheet designations, but where a portion of a basic sheet only appeared on one combined sheet its number was left unqualified in that sheet's designation. These distinctions are also indicated on my map.

Part II of 'An Irish miscellany'
Sheetlines 30, April 1991

Combined sheets of the Irish one-inch coloured edition

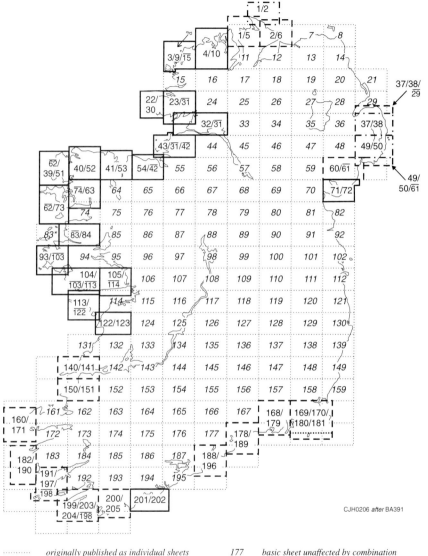

CJH0206 *after* BA391

··········	originally published as individual sheets
——	originally published as combined sheets
—·—·—	as above, but later superseded
— — —	individual sheets subsequently combined
············	sheet line as altered in 1908

177	basic sheet unaffected by combination
2/6	combined sheet
$\overline{198}$	included as 'Part of' in sheet designation '&' is to be understood in designations
*	Sheet 83, the only basic sheet to remain published seaward of combined sheets

Delamere makes a comeback

Delamere Forest is a name known to numbers of Charles Close members, some indeed having probably gleaned it from my writings. So it was that I almost shot from my chair when the name suddenly appeared on my television screen during September.

I have an interest in transport matters second only to that in cartographic matters, and so I regularly watch Ceefax pages 432 to 436 detailing cancellations and other problems on the national rail and London transport systems. It was 434 that recorded another of the unprecedented number of landslips which have occurred on our railways this year, with its heavy though sometimes localised downpours during otherwise dry periods. This one had closed 'the railway through Delamere Forest', causing diversion of trains between Chester and Northwich for a couple of days.

For the uninitiated, Delamere Forest was the original name of an early triangulation station, generally known in later years simply as Delamere, which was the origin (co-ordinate zero) for all the one-inch maps of England and Wales from the second generation of the Old Series, north of the Preston-Hull line, right through to the Popular Edition, plus the Popular Edition of Scotland and most smaller scale maps of the same period. Situated on Pale Heights, a hill to the south of the main forested area, it had been occupied as a primary station in 1803 and 1842 and was the effective hub of Ordnance Survey mapping for the best part of a century. Its accepted geographical position from Clarke[1] is 53° 13' 17".274 N, 2° 41' 03".562 W on the Airy spheroid.

Sheetlines 65, December 2002

Brian explained the profound significance of the NW corner of Old Series sheet 86 and its relationship with the Preston-Hull line and Delamere in a letter on 3 September 1995:

I term the southern sheet lines of [Old Series] 91-94 to be the Preston-Hull line proper, being an exact northing line of Cassini on Delamere running through the north-west corner of Old Series sheet 86, the point which required the smallest area of overlap with previously published mapping. The decision to produce what I have called the Joint Series sheets [91-110] on the Delamere Cassini must have been taken before the top row of the previous generation of Old Series sheets, 87-90, had been constructed, because I have confirmed that their northern borders do conform precisely to the true Preston-Hull line, and bear no particular relation to their southern borders. In particular the northern borders of 87-89 were all made exactly 27 inches/miles across, and the next row up was set out so that 92, 93, 88, 87 all met at a single point, and 91SW, 91SE, 90, 89 at another. If you think all this through you should see that the entire sheet line systems of the New Series, Third and Popular Editions of England and Wales and the Popular of Scotland, not to mention other scales, depend on the north-west corner of Old Series 86, and the co-ordinate systems depend on the figures given to that corner on Delamere Cassini [376,470 feet East, 207,340 feet North], having itself been laid down from Greenwich by, shall we say, "early methods".

---

[1] Clarke, A R, *Account of the observations and calculations, of the principal triangulation; and of the figure, dimensions and mean specific gravity, of the earth as derived therefrom*, London: Ordnance Trigonometrical Survey of Great Britain and Ireland, 1858.

Tardebigge – from Mastermind to minimal mapping

A question about Tardebigge in the last series of Mastermind sent me to my one-inch maps, with somewhat shattering results. It had not occurred to me that there was extensive generalisation and omission of locks on our former prime national series.

Tardebigge Flight consists of thirty locks with a total rise of 217 feet on the Worcester and Birmingham Canal to the south-east of Bromsgrove. It effects almost half of the canal's overall ascent from the River Severn to the Birmingham Level, and is quoted as the longest flight of locks in the British Isles. To be precise it has the greatest number of locks; other flights have marginally greater horizontal extent or total rise. I repeat it has thirty locks, but my one-inch map showed only eleven, eleven lock symbols that is, and a glance at other series and other flights suggested a more detailed study. Immediately I have to amend my statement – the number of lock symbols shown along the Tardebigge Flight on Seventh Series overlapping sheets 130 and 131 varies from ten to twelve on their different states.

My study examined the depiction of Tardebigge Flight and the contiguous Stoke Flight through the various series of one-inch maps, using 1:10,000 sheet SO96NE as a control, and used bridge to bridge sections for a convenient breakdown of the 3¼ mile pair of flights. The table gives the number of lock symbols shown on each section of the flights on the various series. For the longer lasting series two or more states were seen, but only the Seventh had any variation between states.

The Old Series omits several locks close by bridges, which are difficult to map accurately, and one other. However, the positioning of those it does include is not always accurate. The lock symbol on this series is smaller than on any later series and is about true to scale lengthwise. The New Series clearly gives the best overall depiction of the flights, especially of Tardebigge Flight where only one lock is omitted, that by bridge 54 where a bend in the lane imposes extra difficulty. However the Stoke Flight suffers badly by contrast; the bottom lock hard by bridge 44 at Stoke Wharf loses one V from its symbol and two locks are lost from section 45-46 which the sheet line now divides, but the top lock has migrated from its true place above bridge 47 to appear above bridge 48, where there never was a lock. This lock in fact was never mapped in its correct place throughout the life of the one-inch; it lies close above the only skew bridge on the flights. The lock symbol on the New Series and its derivatives through to the Sixth (Popular Edition style) is equivalent to about 140 feet on the ground.

The Revised New Series (NS-2-O) was the first to use the familiar bridge symbol with wing abutments, and its wider size resulted in three more locks being reduced to what I have called half symbols (see note *a*), whilst Stoke Bottom Lock was wholly or largely obliterated (see note *f*). Inexpert deletion of bridge 53 on the Third Edition small sheet series (NS-3-O) left a half symbol in the middle of a section. Older methods of production of separate colour plates led to the situation in note *g*, and to a tendency in some places for the lock symbols to wander onto the towpath or the fields on the other side, but the Popular Edition did see the demise of the misplaced Stoke Top Lock by bridge 48.

These various factors resulted in a gradual decline from 33 to 25 in the number of lock symbols shown along the two flights on the New Series and derivatives, but the next step to the Seventh Series virtually halved the number in a single swipe. On this occasion the Stoke Flight fared better with an increase from one lock symbol to three to indicate the six lock flight, while the Tardebigge Flight was served a decrease from 24 symbols to ten to map its

Number of lock symbols shown on each section

One-inch series (details below)

Bridges	No. of locks	1	2	2R	3SS	3LS	4	6	7p	7q	7r
44 - 45	1[k]	1	1[a]	0[f]	0[f]	0	0	0	0	0	0
45 - 46	3[k]	3	1[l]	1[l]	1[l]	0[g]	1[l]	1	2	2	2
46 - 47	1[k]	0	1	1[a]	1[a]	0	0	0	1	1	1
47 - 48	1[k]	0	0	0	0	0	0	0	0	0	0
above 48	0	-	1[e]	1[ae]	1[ae]	1[ae]	-	-	-	-	-
Stoke Flight	6	4	4[ae]	3[bef]	3[bef]	3[bef]	1[aeg]	1	3	3	3
48 - 49	2	2	2	2	2	2	2	2	1	1	1
49 - 50	5[k]	4	5	5	5	5	4	4	1	1	1
50 - 51	5[k]	3	5	5	5	5	5[b]	4	2	2	2
51 - 52	4[k]	3	4	4[a]	4[a]	4[a]	3	2	1	2	2
52 - 53	2[k]	1	2	2	}5[am]		4[agl]	4[l]	4	2	2
53 - 54	3[k]	2	3	3						2	2
54 - 55	8[k]	8	7	7	7	7	7	7	3	4	3
55 - 56	0	-	-	-	-	-	-	-	-	-	-
56 - TTM	1	1	1	1	1	1	1	1	0[n]	0[n]	0[n]
Tardebigge Flight	30	24	29	29[a]	29[b]	28[bg]	26[b]	24	10	12	11
Total: two flights	36	28	33[ae]	32[cef]	32[def]	29[ceh]	27[b]	25	13	15	14

(In rows 52 – 53 and 53 – 54 the columns 3SS, 3LS, 4, 6, 7p, 7q, 7r are shown combined: }5[am], 4[agl], 4[l], 2, 2, 2.)

Notes:

a   includes one lock with half symbol (one V only)
b   two locks as note a
c   three locks as note a
d   four locks as note a
e   one lock sited above wrong bridge
f   excludes mark possibly remnant of half symbol
g   excludes a narrowing in the blue, with black symbol missing

h   excludes two places as note g
k   one lock close by lower bridge
l   section crossed by sheet line
m   Bridge 53 demolished
n   Bridge 56 and Tardebigge Top Lock obscured by name

TTM Tardebigge Tunnel mouth

Series code	Description		Sheet numbers ‡
1	Old Series		54NW
2	New Series	NS-1-O ‡	182, 183
2R	Revised New Series	NS-2-O ‡	182, 183
3SS	Third Edition small sheet series	NS-3-O ‡	182, 183
3LS	Third Edition, Large Sheet Series (coloured)		72, 82
4	Popular Edition (including War Revision)		72, 81, 82
6	New Popular Edition (Popular Edition style)		130, 131
7p	Seventh Series		130 (A, A/), 131 (A, A//)
7q	Seventh Series		130 (A//, A///*)
7r	Seventh Series		130 (B), 131 (B/*, C, C/*)

† Richard Oliver's series code

‡ Series 2 - 4 sheets abutting; 6 and 7 overlapping

thirty locks. Particularly unfortunate was the total obliteration of the singleton Tardebigge Top Lock by the name of Tardebigge village. As mentioned above, this series differed between states in the number of locks shown, and ended with fourteen on each of the overlapping sheets. The main problem on the Seventh Series was the adoption of a bolder lock symbol equivalent to about 230 feet on the ground, which made it impossible to show successive locks spaced about every 320 feet on the steepest parts.

I next took a look at the depiction of other major flights of over twenty locks on late states of the Seventh Series, thus:-

Flight	*Canal*	*No. of locks*	*Seventh Series*		*% locks on map*
			locks	*sheets*	
Tardebigge	Worcester & Birmingham	30	11	130, 131	37
Devizes	Kennet & Avon	29	23 [x]	167	79
Wigan	Leeds & Liverpool	23	14	100, 101	61
Hatton	Warwick & Birmingham [y]	21	15	131	71
Wolverhampton	Birmingham Main Line	21	8	119, 130/1	38

[x] locks and disused ponds   [y] constituent of Grand Union

The good showing of the Devizes Flight here is due to the relatively accurate depiction of the disused ponds on the Caen Hill section, whereas it would not have been possible to include more than fifteen lock symbols at the very most had this ascent still been in use (at the time of revision). On both the Hatton and Wigan Flights the locks are more spread out, allowing a greater proportion to be mapped, but the lower end of Wigan shares with Wolverhampton a congested urban passage which militates against insertion of lock symbols in Seventh Series cartography. Thus the final picture is that the number of lock symbols depicting lock flights on Seventh Series maps is roundly half the actual number of locks concerned.

I have left the good news until last. I had envisaged this article ending with a blast against the modern tendency, only partly due to modern draughting methods, towards over-generalisation, which makes the map much less of an immediate picture of the country. I do not abate my views on this, but I must give credit where it is due. To complete the picture I looked at the 1:50,000 cover of the pair of flights on sheets 139 and 150. The 1:50,000 First Series, being derived directly from the one-inch Seventh, naturally shows the same fourteen lock symbols as appear finally on that series. But the newly drawn Second Series, with a new slim-line single V symbol, manages what the entire one-inch series never did; it maps every one of the 36 locks on the Stoke and Tardebigge Flights at the same time. Nevertheless, having examined the cartography of these flights on the one-inch maps in minute detail, I have to record one error on the new scale – the lock close by bridge 50 lies immediately south of the grid line which divides the two sheets, but the symbol appears on the northern sheet, 139.

Letters to the editor and other short pieces

The longevity of the Third Edition small sheet series of the one-inch map of England and Wales

Richard Oliver mentioned to me on a recent visit that he had seen printings of the above series dated as late as 1942. This reminded me of an incident nearly fifty years ago which may throw some light on the longevity of this series.

At a period during the Second World War when the civil defence services in eastern England were becoming increasingly involved with Allied bombers which had crashed when limping home from distant raids, the Senior Warden in my mother's Fens-edge village enquired from his superiors "Exactly what area are we responsible for?" He showed me what he was sent in response, a cutting from a black-and-white Third Edition map with the boundaries of the parish inked over. Here surely we have a reason, possibly the only reason, for the continued use of this series in administrative circles – a one-inch map with the parish boundaries still shown thereon. Although the boundaries would by then have been considerably out of date, the majority of the alterations would have been amalgamations of parishes, keeping the requisite boundaries intact with some superfluous ones to be ignored.

Sheetlines 27, April 1990

Parish lists of the 1930s

Brian Adams draws our attention to lists of parishes with the relevant six-inch sheet numbers published by the Ordnance Survey in the 1930s. An extract from that for Bedfordshire, dated 1934, appears below, and a collection of them for England and Wales has printing dates varying from 1932 (Oxfordshire and Warwickshire) to March 1939 (Yorkshire, West Riding). It would be interesting to know why these lists were produced; were they a replacement for the more elaborate *Catalogues*, of which the last had appeared in 1920, or were they to serve some administrative purpose?

Hockliffe .... 28 N.E., S.E.; 29 N.W., S.W.	Toddington .. 25 S.W., S.E.; 29 N.W., N.E., S.W., S.E.
Houghton Conquest...................... 16 S.E.; 21 N.E.	Totternhoe .... 28 S.E.; 29 S.W.; 31 N.E.;
Houghton Regis......... 29 S.W., S.E.; 32 N.W., N.E.	32 N.W., S.W., S.E.
Hulcote and Salford................. 20 N.W., N.E., S.E.	Turvey .... 10 N.E., S.E.; 11 N.W., S.W.
Husborne Crawley.........20 S.E.; 24 N.E.; 25 N.W.	Upper Stondon ....26 N.E.
Hyde .... 33 N.W., N.E., S.W., S.E.; 35 N.W., N.E.	Westoning ....25 N.W., N.E., S.W., S.E.
Kempston .... 11 S.E.; 16 N.W., N.E.	Whipsnade............................ 32 S.W., S.E.; 34 N.E.
Kempston Rural................ 11 S.W.; 15 N.E., S.E.;	Wilden............................8 S.W., S.E.; 12 N.W., N.E.
16 N.W., N.E., S.W., S.E.	Willington ....12 S.W., S.E.; 17 N.W., N.E.
Kensworth .... 82 N.W., N.E., S.W., S.E.	Wilshamstead ....16 N.E., S.E.; 17 N.W., S.W.;
	21 N.E.; 22 N.W.

Sheetlines 33, April 1992

Visit to Royal Engineers Museum, 2 May 1992

For a former unpaid part-time Sapper[1] an afternoon at the Royal Engineers' Museum, Chatham[2], was an exhilarating experience, the icing on the cake being a talk by our then chairman-to-be, Yolande Hodson. My afternoon began unusually enough when, on the way from Gillingham station, a stranger asked me the way to the museum and then explained 'I am the present Earl Kitchener'. The formal proceedings began with an introductory talk by the Director of the Museum, Colonel Gerald Napier, setting the scene for R.E. surveying within the wide context of the Corps' history and responsibilities. Yo Hodson, who is also Hon. Secretary of the Palestine Exploration Fund, then gave her illustrated talk on 'The survey of Western Palestine 1871-1878' sponsored by that fund and executed by the Royal Engineers, a comprehensive operation which is unfortunately less well-known than the Ordnance Survey's smaller survey of Jerusalem.

This part of the proceedings was concluded by Rev. Raymond Goodburn, who delivered a travelogue on the Biblical sites today. Some of these sites situated on commanding heights had been adopted by the 1870s survey as primary triangulation stations, and were illustrated by both speakers, but from different aspects. The party then visited the museum's special exhibition 'Royal Engineer surveys in the Holy Land', after which they were treated to a conducted tour through the whole museum by Col Napier. The final act was a sumptuous tea prepared by Mrs Sharren Jones of the museum staff, during which well-deserved tributes were paid to all those who had contributed to a most successful afternoon.

Sheetlines 34, September 1992

Puzzle corner

Puzzles contributed by and for members have been used to fill the odd spare corners of Sheetlines over the years. At least two were posed by Brian Adams. In Sheetlines 25, to accompany the first part of '198 years and 153 meridians, 152 defunct', he asked,

Which county origin has the same name as a well-known TV presenter whose home is in that county?

and in Sheetlines 32,

Which one-inch Popular Edition sheet of England and Wales is not filled to the neat line?

The answers will be found overleaf.

---

[1] R.E. Unit, Cambridge University Senior Training Corps/8th Cambs Home Guard.
[2] Those with good editions of maps, showing parish boundaries, will see that it is actually in Gillingham.

Was it me, or the Ordnance Survey?

Nick Millea brought to the attention of readers of Cartographiti 70 *a St George's Day card published by Hallmark Cards, showing a map of Great Britain that highlighted England, with Scotland, the Isle of Man and Wales in the background. The Anglo-Welsh border was depicted running east of Shropshire and Herefordshire! This oddity led Brian to comment; we include his contribution to* Cartographiti 71 *(June 2004) by kind permission of the editor.*

The Cartographic Curiosity in *Cartographiti 70* poses a number of questions. Is it a case of territorial aggrandisement as Nick Millea suggests, or is it rather a matter of saintly allegiance? A pause for thought suggests not, for a dragon-slaying saint will surely be especially venerated along the Marches. So was it then Llewelyn's dream, or Offa's nightmare?

But wait; does not that line seem familiar? Of course, I have drawn it myself, several times, and so has the Ordnance Survey in its earlier, non-National Grid days. In fact the OS produced a whole succession of maps, many in *Descriptions* booklets, over a period of fifty-five years, categorised by me as their most inaccurate maps (see page 59). But the line in question was not among the errors on those maps, forming as it does the boundary between the fifteen 'central counties' whose County Series plans were constructed on the origin of Dunnose and the twelve Welsh and Marcher counties constructed on Llangeinor, plus a portion at the top between the Cyrn-y-Brain counties and Cheshire on Nantwich Church Tower.

My own first involvement with this line was in the production of a Hydrographic Department restricted issue booklet in 1946. But it was not for another 43 years that I drew it under my own name, in *Sheetlines 26* and in a number of subsequent versions, the latest in the same week that I write this piece. So we are left with the biggest question of all, where on earth did Hallmark Cards UK get the line from?, and who failed to realise that the line did not delimit the area under the patronage of St. George?

Puzzle corner – the answers

The reference in the first question is to Rippon Tor, county origin of Devon, the home county of Angela Rippon. The second answer is England and Wales Popular Edition sheet 6; the small portion of Scotland in the north-west corner is left blank.

Another late running railway

Another instance of 'Railway in course of construction' seriously out of time could be found some eight miles north of the one reported by Aidan de la Mare in *Sheetlines 62*; eight miles as the crow flies that is, but fifteen by river.

On 1:25,000 sheet 20/46 Provisional Edition, published in 1946, the annotation appears twice along the course of the line linking Bere Alston to a point near Albaston by a tortuous inverted-S crossing the Tamar at Calstock, although the line had in fact been opened in 1908. The annotations had been transferred from six-inch sheet Cornwall XXX SW, second edition 1907, still current in 1946.

This line had been built by the Plymouth, Devonport and South Western Junction Railway to link the L&SWR main line to the previously detached East Cornwall Mineral Railway. However, the latter had been a narrow 3'6" gauge line and it was converted to standard gauge from Albaston to the terminus at Callington Road, previously named Kelly Bray and later Callington, although it still remained more than a mile from that village.

Sheetlines 63, April 2002

Brian also wrote letters to the railway press (e.g. Modern Transport, *19 April 1947, 16 August 1947); perhaps his most significant contribution was to* The Railway Magazine *in November 1958, in which he gave a practical demonstration of the value of county co-ordinate data to determine a point of debate in the railway press at the time: where was the most westerly railhead in Europe – at Valentia or Dingle?*

… Valentia Harbour has been the most westerly railhead in Europe ever since the line was opened there in 1893, although Dingle held the distinction for 2½ years previously. Since the railheads lay almost due north-south, it is necessary in order to establish which was more westerly to go to the accuracy obtainable on the Ordnance Survey 6-inch map, from which the respective railheads may be accurately scaled off. The buffer stops at Valentia Harbour will be found to be 11,564 feet east, 7,715 feet north from the south-west corner of Kerry sheet 79, which gives them Kerry county co-ordinates of 129,636 feet west, 124,085 feet south; making allowance for a county twist of 3.11 seconds these convert to a geographical position of latitude 51° 55' 44.8" north, 10° 16' 40.1" west. Similarly the ultimate railhead of the Tralee & Dingle Railway at the root of Dingle pier was 12,855 feet east, 204 feet north of the south-west corner of Kerry 43 or 128,345 feet west, 47,596 feet south on Tralee Church Spire, which gives latitude 52° 08' 19.2" north, 10° 16' 29.1" west. Hence the Valentia Harbour buffer stops were 11.0 seconds of longitude or 690 feet west of the meridian of the Dingle Harbour railhead, and the claim of Valentia Harbour to have had the westernmost rails in Europe ever since they were laid down is vindicated.[1]

He wrote again in October 1964 (in the event, unpublished), giving the lie to the myth that the sun would shine right through Box Tunnel at sunrise only on I K Brunel's birthday, 9 April. Brian's own computations supported those of the Superintendent of H M Nautical Almanac Office, and confirmed that it could do so on 7 April, possibly 6 or 8 April in some years, but never later. What seems to have been ignored, perhaps even by Brunel himself, was the effect of the sun's light being refracted in the atmosphere. Without this factor, the sun would indeed have shone directly through the tunnel on 9 April!

---

[1] This extract printed by kind permission of the editor of *The Railway Magazine*.

Other publications by or about Brian Adams

F G B Atkinson, B W Adams, *London's North Western electric: a jubilee history*, Sidcup: Electric Railway Society, 1962. pp. 48, with 16 illustrations on 4 plates, lists, map and plan. [E R S monograph series 1] [Ottley 6391].

Appendix 7: Geographical & Cassini co-ordinates, in Yolande Hodson, *Popular maps: the Ordnance Survey Popular Edition one-inch map of England and Wales 1919-1926*, London: The Charles Close Society, 1999, ISBN 1 870598 15 6, [Brian contributed the fundamental data, checked the Cassini co-ordinates and recalculated the geographical co-ordinates transcribed by Yolande from Ordnance Survey sources].

Appendix 7: Sheet line co-ordinates, in Roger Hellyer and Richard Oliver, *A guide to the Ordnance Survey one-inch Third Edition maps, in colour: England and Wales, Scotland, Ireland*, London: The Charles Close Society, 2004, ISBN 1 870598 21 0.

HD 364, Ordnance Survey sheet line co-ordinate data for six inch and 1:2,500 sheets, Hydrographic Department of the Admiralty, Restricted Issue 1946 [issued anonymously].

Marginalia, *Sheetlines 37* (1993), 2.

The Hodson award, *Sheetlines 46* (1996), 2: the announcement that the initial award was made to Brian Adams, in recognition of his work on the mathematical aspects of the cartography of Ordnance Survey maps.

Review by Alan Godfrey of *Ordnance Survey of Great Britain: Scotland – Indexes to the 1/2500 and 6 scale maps*, Kerry: David Archer, 1993; in *Sheetlines 37* (1993), 6.

Peter Haigh, 'The bases of the Ordnance Survey', *Sheetlines 62* (2001), 42-56, in which the author acknowledged Brian's assistance in contributing fundamental data.

Letter in response to question "How many islands are there in the British Isles?" in 'Notes & Queries' column in *The Guardian*, 25 June 1990. The question had been posed on 4 June.

Letter in response to question "Exactly how many islands make up the British Isles?" in 'Answers to correspondents' column in *Daily Mail*, 18 July 1995 (from which a crucial line was omitted, detailing individual totals for Scotland of 7,260, Wales 250, Northern Ireland 120, Republic of Ireland 1,890).

'?How long is a piece of string? or how long are ten cables' – a chart of the waters surrounding the British Isles, prepared (circa 2000) by Brian Adams (Parsons Green Hydrographic & Geodetic Consultancy), to demonstrate that a sea mile (= 10 cables = 1 minute of latitude) is of variable length, while an international nautical mile (= 1852 metres) is of fixed length.

'Adoption of Greenwich Meridian', apparently an internal Hydrographic Department paper completed on 14 August 1973 recording the history of the meridian since 1676, and noting the existence of its three distinct alignments.

Rockall – information summary (H 4500/69), Taunton: Hydrographic Department (Marine Sciences Branch One), August 1969; together with *Rockall – information summary – addendum 1973*, November 1973, both accredited to Brian Adams.

'Brian'

David Archer

I have no recollection of ever being introduced to Brian, and like most people, I simply became aware of him. Brian was a great attender and went to most meetings. He was also very distinctive: short in stature, with a beautifully shiny and polished head. Slightly pointed, with a neat ring of hair just above the ears. In all other matters, Brian was a hairy man, with thick dark eyebrows and a strong growth peeking out of his cuffs and shirt neck. Given his character, I am positive that he had hairy feet as well. One could not help but like him. He always had an ETA for meetings and would appear and hesitate at the door, often with a small shopping bag, before launching himself towards a friend. Being ever so slightly stooped, with his head forward, he seemed to almost overbalance and progress across the floor in a gliding shuffle. Brian had arrived. On such occasions, it was exceedingly difficult to hold a conversation, as he was also a very quiet person. He whispered. In recent years he seemed to turn the volume even lower, so low that if one wanted verbal information, one had to use the telephone in order to hear what he said.

Others will tell of Brian's brilliance, of his mathematical prowess and its application to matters cartographical, but I remember him especially for the fun. Not ho-ho-ho joke-telling fun, though he had an acute sense of humour, but for the little things that made Brian a fun person, a character but not an eccentric. Most of the fun things that I remember of Brian, arose from his devotion to detail and accuracy. The accuracy of a mathematician, and the meticulousness of someone who knows what is required in order to produce a result that would attract full marks. Brian never aimed for less than full marks.

In 1992, when I was Secretary, the committee decided to ask Brian to talk at the next AGM, so I approached him in good time, shall we say in October for the following May. Yes, he would speak, but it would need a lot of preparation, so that it would have to be the AGM after next. Just about enough time. He then went underground on the matter. Again, I cannot give a date, but it would have been months before the AGM, probably December or January; Brian rang and wanted to know the room number in Birkbeck that we had booked for the great day. I could not give one, as I think there was the possibility of one of several due to exams. He was most put out by this, and could not go any further with his preparations until he had visited the room in which he was to speak. So I told him that it would be the big double room on the third floor, that we had used before. I never found out why, but he went there a couple of times and the preparations continued. As May got nearer, so the telephone calls became more frequent. Which end of the room would he be speaking from? Where exactly would he have to speak from, could an easel be provided for the flip chart he was making, and how high would it be? And other minor points that I now forget, but things which all too often can mar a talk if not taken into account. Brian was planning down to the last detail within details.

In the week before the AGM, it was agreed that I would collect his flip chart, his long wooden pointer and various other things on the Friday, and take them to Birkbeck on the Saturday morning for Brian to set up. I then spent the whole of Thursday, trying unsuccessfully, to arrange for some sort of amplification system in the room. A double room, and eighty people for Brian to whisper to. Why had I not thought of this before? We arrived early and started to prepare the room, expecting Brian to be there already, but he did not appear. Coffee was almost finished and still no speaker had arrived. These were the days before mobile phones, not that he would have had one (or did he? Full of surprises). At last, our man appeared at the door, and although extremely flustered, went straight to the front and calmed himself by putting his things in order. I asked if he was all right, and he said that a tube train had not run according to the timetable. No more was said, he was furious with London Transport. The whole morning had been planned around the Underground timetable, which had failed. Until then, despite having been raised in London, I had never even considered that tubes had timetables, they just arrived and one left home in plenty of time to get from A to B. Not Brian, he had researched the times, (probably undertaken a dummy run the previous Saturday), and they had failed him on his big day. But our speaker did not fail the society, and sitting at the very back of the room, I heard most of what he said.

Anything that involved preparation, such as answering questions, writing papers, or speaking at the AGM became a "job", and was added to the pile, to be dealt with when its turn came. Big jobs and little jobs were inter-filed and seldom could one speed their rise to the top. One just asked for something and forgot about it until it turned up, and one was more than delighted with the result. Brian wrote two pieces for inclusion in our publications, and after agreeing to do them, we could get nothing out of him until he sent the final text. No drafts, just final polished and word perfect text. This obviously caused us problems in that we had no idea how long the pieces would be, and hence could not finalise all sorts of things such as page numbers, indexes or even how long the book would be and hence get final quotations for printing. We just left some space, and were prepared to have a longer book if necessary, as what would arrive could not be cut. Jobs inevitably piled up, but he could never be rushed, and would worry if pressured. With any job, Brian just went his own way; as he knew what was needed, and it had to be done properly, so there was no point in discussing it with others.

When Landmark Information Group were getting things off the ground, they approached me, seeking publications that were of a mathematical nature, in order to help with the complex calculations needed for their vast set up. I immediately put them on to Brian, as being the only person who could undertake the calculations needed. Being Brian, he was very wary at first, and wanted to know the ins and outs of what they were doing. I could not tell him, so they invited him down to see the set-up and discuss things. As a result, he agreed terms and was installed as a consultant. Whilst being very pleased with the recognition the title gave, it caused him no end of amusement. He rose to the occasion by announcing the formation of PG consultancy, with nicely printed compliment slips, sent from the corporate headquarters in Parsons Green. Such was his hold on the market that he subsequently became the cartographical guru for Sitescope and Timeline Maps. Brian was justifiably proud of his technical writings and proud to have won the Hodson Award for them. But he also expected society members to have read and remembered what he had written. Although his writings on quite complex matters are always wonderfully clear, sometimes the facts do not remain in the reader's memory. The reader being me. In this, I lived in fear of his suddenly testing me on the difference between an origin and a meridian, or of having to explain in twenty words, how grids tie in with projections.

Brian liked a nice organised life and planned things accordingly. Hence, when the London telephone numbers were changed, he sent out slips with his new number, and an instruction not to telephone at certain times in the evening, as he did not want to be interrupted, the certain times being those of University Challenge and Mastermind. Similarly, although Brian went everywhere by public transport, (and would travel long distances to deliver manuscripts, not trusting the post with such important documents), he was sympathetic to the problems of car owners. Living in central London, with bumper to bumper street parking, he was exceedingly organised to receive visitors. If he knew you were coming by car, he would keep a space for you, and had two red and white plastic cones with a plank to go across them, which he would put in the road outside his flat. When the society visited the National Railway Museum, we were shown around the library and given free range to browse. Brian disappeared into the stacks, eventually returning with a mischievous twinkle in his eye, plus a copy of a small volume on railway history, that he had co-written, and asked the librarian if the museum would like him to sign it. Great fun, and a good tale, but only one of many. If he were to read this piece, he would surely point out the slight exaggerations and small errors of memory that creep in with time. But he would also, I am sure, be pleased that his thoroughness was appreciated. As I have stressed, Brian thought that if you did something, it had to be done spot on, and this involved lots of research, preparation and polishing.

Whenever he telephoned, Member 41 would always start the conversation "Brian here, David". Not that one needed any identification, given the ever so distinctive voice and slightly nasal delivery. Amongst society members, by itself, the word Brian has only ever referred to one person, and I tend to think that he quite liked the name. His opening remarks to the 1994 AGM mention "the blue plaque (not yet installed) 'Brian Was Born Here'". He certainly had great fun in the 1980s when he saw Bryan Adams splattered across newspaper headlines and record shop displays, "but it's spelt differently" he would note. And I feel certain, that if he could have obtained a copy, somewhere amongst his most treasured maps, lurks a poster for the Monty Python film celebrating, *The life of Brian*.

Index